HEAT EXCHANGERS

Holger Martin

Institut für Thermische Verfahrenstechnik
Universität Karlsruhe, Germany

●HEMISPHERE PUBLISHING CORPORATION
A member of the Taylor & Francis Group

Washington   Philadelphia   London

HEAT EXCHANGERS

1 2 3 4 5 6 7 8 9 0   B R B R   9 8 7 6 5 4 3 2

Originally published as Wärmeübertrager by Georg Thieme Verlag Stuttgart, New York, 1988

Translated by the author and Vijay R. Raghavan.

Cover design by Kathleen Ernst.
A CIP catalog record for this book is available from the British Library.

Library of Congress Cataloging-in-Publication Data

Martin, Holger, date.
   [Wärmeübertrager. English]
   Heat exchangers / Holger Martin.
     p. cm.
   Includes bibliographical references and index.

   1. Heat exchangers. I. Title.
TJ263.M365   1992
621.402′5—dc20   91-29040
ISBN 1-56032-119-9   CIP

CONTENTS

Chapter 3   Examples in Heat Exchanger Design   115

PREFACE TO THE ENGLISH EDITION

The production of an English translation of my book *Wärmeübertrager* was suggested by Jerry Taborek, formerly director of Heat Transfer Research Inc. (HTRI) in Alhambra, California. He was a visiting scientist in our laboratory in Karlsruhe in 1986–1987 as a Humboldt fellow when I was writing the German text, which was published in Sept. 1988 by Georg Thieme Verlag Stuttgart, New York.

In February and March 1989, I was at the Indian Institute of Technology (IIT) in Madras as a guest professor and used material from the book for my lectures there. Vijay R. Raghavan, Professor and Head of the Heat Transfer and Thermal Power Laboratory at IIT Madras, kind and generous host in those days, dear friend in the meantime, has shown keen interest in the contents of these lectures. He even expressed a desire to translate my book.

In October 1989, when the Hemisphere Publishing Corporation actually approached the German publisher for a license to produce an English edition, I wrote to Prof. Raghavan to inquire if he would assist me in the translation, to which he readily agreed. The first draft of my very crude translation took the form of some 11 notebooks, which were sent to Madras between Dec. 1989 and Feb. 1990. He sent me back the corrected notebooks one by one, with my crude draft turned into readable clear English. In the process of translation, the book has also undergone a technical review, so much so that the English version has turned out to be practically an improved second edition.

I have also incorporated results from recent papers that appeared after the publication of the original (references [B3a], [B3b], [R4]) and from an old paper which has been brought to my attention only recently (reference [K4]; see also the new Appendix C).

The notation used in this book is essentially the same as that recommended for the International Heat Transfer Conferences and used in the Heat Exchanger Design Handbook (HEDH) [H3] since 1983. Hopefully, English-speaking heat transfer engineers have become accustomed in the meantime to find *heat transfer coefficients* denoted by α (lowercase Greek *alpha*) in place of *h;* one good reason for this change is the internationally well-established use of *h* for specific enthalpy. HEDH, nevertheless, has retained the traditional (English) notation *U* for overall heat transfer coefficients, in spite of its parallel use for internal energy. In this book the symbol for *the overall heat transfer coefficient* is *k,* which is also recommended internationally as an alternative to *U,* but not widely used so far, probably because *k* has been conventionally used for the *thermal conductivity,* now internationally denoted by a λ (lowercase Greek *lambda*). In case of doubt, a look on the list of symbols, page 197, should help avoid confusion.

In addition to expressing my deep gratitude to all those who encouraged, suggested, and produced this English edition, I would like to express my hope that the book might be useful for those studying and for those professionally working in the field of heat transfer and heat exchanger design.

Karlsruhe, Winter 1990   *Holger Martin*

PREFACE TO THE GERMAN EDITION

For many engineers, *Wärmeübertrager* (literally "heat transmitter") in place of *Wärmeaustauscher* ("heat exchanger") may still be a somewhat unfamiliar term for the appliance in which heat is transmitted steadily from one medium having a higher entrance temperature to another medium with a lower entrance temperature. Thermodynamicists such as Ernst Schmidt, had already used the more correct expression (i.e., *Wärmeübertrager*) in preference to the currently used one (i.e., *Wärmeaustauscher*) [S5]. Now that the VDI-Wärmeatlas [V1] too has replaced *Wärmeaustauscher* by *Wärmeübertrager* since 1984, it seems appropriate to use this term generally in engineering education.

The present book is addressed to students of engineering and science, especially in the fields of technical chemistry, chemical and process engineering, mechanical engineering, and physics. Knowledge in mathematics, thermodynamics, heat and mass transfer, and fluid dynamics, as usually obtained in universities or institutes of technology after three years of studies, are thought to be a prerequisite.

The subject of the book is not heat transfer but its application to the calculation of temperature profiles, especially the outlet temperatures of both media and the transfer performance of heat exchangers.

In Chapter 1, three examples illustrate in detail how to apply the fundamentals of thermodynamics, heat transfer, and fluid dynamics for a systematic analysis of the phenomena in heat exchangers. The systematic procedure for the solution of problems in this field is set out in the form of a comprehensive scheme.

Chapter 2 is dedicated to the influence of flow configuration on the performance of heat exchangers. Here the equations to calculate mean temperature difference and efficiency for stirred tank, parallel, counter- and crossflow, and their combinations are derived and put together in a very compact way. In some cases, short computer programs are given to evaluate more complicated formulas or algorithms. Therefore, the book should also be useful to practicing engineers as a reference for these relationships. It is so written as to enable one to work through the contents alone with appropriate preparatory training.

The fully worked-out examples in Chapter 3 are intended to show the application of the fundamentals to thermal and hydraulic design, i.e., sizing of heat exchangers. Mechanical design, with choice of material and calculation of strength according to relevant construction codes, has not been included. The latter is the subject of the course *"Konstruktiver Apparatebau,"* for which a similar book would be desirable.

The present book was developed as a text on the basis of the course *"Kalorische Apparate A,"* offered for many years at the University of Karlsruhe by Professor Dr.-Ing. Dr.h.c.INPL Ernst-Ulrich Schlünder, which I have taken over from the winter term of 1986–1987. It was Prof. Schlünder who suggested that I write this book. The entire conception and a majority of the examples are engendered by his ideas. Apart from the elaboration of the hitherto handwritten course notes, my own contribution was restricted to the more recent research results on plate and spiral plate heat exchangers, which are mainly based on the work of my former student Dr.-Ing. Mohamed K. Bassiouny as well as on the compact representation of the most important analytical results on the influence of flow configuration on heat exchanger performance developed at the end of Chapter 2. To the original course contents, I have added the analysis of heat exchangers coupled by a circulating heat carrier in order to assist the reader in comprehending the phenomena in a regenerator. All the numerical examples have been reworked, using the calculation procedures for heat transfer coefficients and friction factors currently recommended in the pertinent handbooks on the subject.

Dr.-Ing. Paul Paikert, director of the Research and Development Department of GEA Luftkühlergesellschaft, provided field data for the design examples on plate and shell-and-tube heat exchangers, which is gratefully acknowledged. I would also like to thank my former colleague Dr.-Ing. Norbert Mollekopf, now with Linde A.G., for information on the design of regenerators and other heat exchangers in flue gas cleaning processes applied in power plants. To my colleague Akad. Dir. Dr.-Ing Volker Gnielinski, I am indebted for his critical inspection of the manuscript and for many a valuable hint on the layout of the book. For the excellent drafting of a majority of the figures, I would like to thank Lothar Eckert and Pedro Garcia. Some of the figures have been obtained courtesy of Linde A.G. (Höllriegelskreuth) and W. Schmidt G.m.b.H. u. Co. K.G. (Bretten). I have myself produced some of the figures, using the graphic software "MacPaint" by Apple Inc.

Finally, I would like to thank Nana very much for carefully transcribing my handwritten notes into neatly typed text stored on a disk. She sacrificed many a weekend for this arduous work.

Karlsruhe, Summer 1988 *Holger Martin*

CHAPTER ONE

ANALYSIS OF SOME STANDARD TYPES OF HEAT EXCHANGERS ON AN ELEMENTARY BASIS

1. STIRRED TANK WITH JACKET

1.1 Description

The stirred tank, or stirred vessel, is one of the simplest and, at the same time, most versatile types of apparatus used in process engineering. In the model shown in Fig. 1.1, the vessel is put together from cylindrical, annular, and spherical shell segments according to structural analysis. The lower part has a double-walled construction with inlet and outlet headers, so that the contents of the vessel may be heated or cooled by a medium flowing through the jacket. In Fig. 1.2, this apparatus is drawn schematically with its most important functional features. The flows of mass and energy entering and leaving the vessel and the jacket are inserted into the sketch as arrows and denoted by symbols, such as $\dot{M}$ for mass flow rate, $\dot{Q}$ for heat flow rate, and $\dot{W}$ for stirrer power, which are, if necessary, identified by subscripts for position, time, or state.

1.2 Formulation of Questions

In the next step, one has to become clear on the questions of which are exactly the unknown and which are the given—or, at least, to-be-fixed-in-advance—quantities. Reasonable questions in connection with heating a liquid in a stirred tank may be , for example,

1

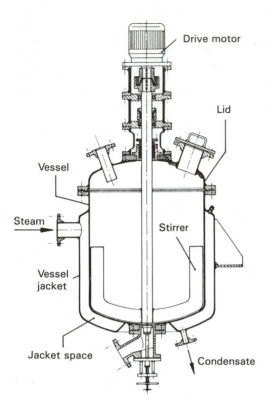

Figure 1.1 A stirred tank with jacket.

a. How does the temperature of the contents of the vessel change with time after the steam inlet valve has been opened?
b. What is the consumption of heating steam?
c. What is the influence of the stirrer speed on the heating process?

To answer these questions, some quantities have to be known or fixed in advance. Only in the course of the analysis will it become apparent that, to answer the question, we need the vapor pressure p_V and, thus, the condensation temperature $T_V = T^*(p_V)$ of the vapor in the jacket, mass M and initial temperature T_I of the liquid in the vessel, type and rotational speed of the stirrer, and other parameters. In any case, it will be useful to list all necessary parameters using unambiguous symbols. Question (a), for example, can be formulated in symbolic writing as

$$T = T(t, \text{parameters})$$

Here T, the unknown (or sought-after) quantity, the temperature of the liquid in the tank, is a function of the time t after the steam inlet valve has been opened and "parameters" contain all other quantities that have to be known *a priori* to calculate the function $T(t)$. In general, before starting the formal symbolic analysis, one has to be clear on the following questions:

- Which is the quantity sought after?
- What does it mainly depend on?
- Which other parameters are needed?

1.3 Application of Physical Laws

To answer the question posed in the previous section, in general, three classes of physical laws are at our disposal: the laws of conservation (of mass, momentum, energy); the laws of equilibrium; and the rate equations (kinetics of transport processes).

Over a space, thought of as the "control volume," one may write a balance for physical quantities that obey a law of conservation. Thus, the mass balance for the steam jacket is

$$\dot{M}_V - \dot{M}_C = \left(\frac{dM}{dt}\right)_{\text{in the steam jacket}} \tag{1.1}$$

If the vapor pressure p_V remains constant and if, by means of a steam trap, the liquid level of condensate in the jacket is also kept constant, then the change of mass (of

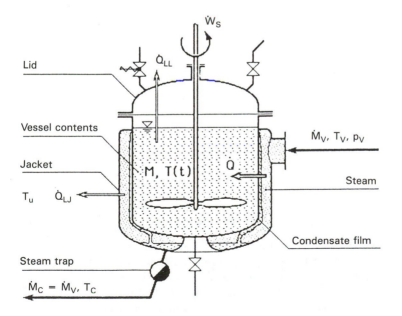

Figure 1.2 Stirred tank: sketch, streams, symbols.

vapor and condensate) in the jacket with time is equal to zero. The amount of vapor flowing in is always equivalent to the amount of condensate led off ($\dot{M}_V = \dot{M}_C$). The corresponding energy balance for the steam jacket reads:

$$\dot{M}_V \Delta h = \dot{Q} + \dot{Q}_{LJ} \tag{1.2}$$

Here $\Delta h = h_V - h_C$, the difference of the specific enthalpies of vapor and condensate, i.e., the enthalpy of vaporization $\Delta h_V(T_V)$ and appropriate additional enthalpy differences in case of steam entering superheated and condensate leaving subcooled. $\dot{Q}$ is the rate of heat transferred from the condensing steam to the contents of the vessel, while $\dot{Q}_{LJ}$ is the heat "loss" from the jacket to the surroundings. In order to answer the question 1.2 (a) for the variation with time of the temperature of the contents of the vessel, we have to regard the contents as a system. In case both inlet and outlet valves remain closed, the energy balance for the liquid contents of the vessel is:

$$\dot{Q} + \dot{W}_S - \dot{Q}_{LL} = \left(\frac{dE}{dt}\right)_{\text{Vessel contents}} \tag{1.3}$$

$\dot{W}_S$ denotes the power transferred by the stirrer and $\dot{Q}_{LL}$ the heat loss from the contents through the lid to the surroundings. In this case (and similarly in many other practical cases in heat exchangers), of the components of the total energy of the system, i.e., the potential, kinetic, and internal energies, only the internal energy U changes. On the other hand, one has to take into account the expansion or contraction of the fluid in the vessel when heated or cooled under constant pressure. Thereby it transfers power by change of volume $\dot{W}_V = p(dV/dt)$ to its surroundings, which has to be subtracted from the left hand side of eq. (1.3)

$$\dot{Q} + \dot{W}_S - \dot{Q}_{LL} - \dot{W}_V = \left(\frac{dU}{dt}\right)_{\text{Contents}} \tag{1.4}$$

By introducing the enthalpy

$$H = U + pV \tag{1.5}$$

the energy balance can be simplified for constant pressure:

$$\dot{Q} + \dot{W}_S - \dot{Q}_{LL} = \left(\frac{dH}{dt}\right)_{\text{Contents}} \quad (p = \text{const}) \tag{1.6}$$

Since the pressure in heat exchangers often remains constant with time, the energy balance can be formulated most conveniently in many cases as in eq. (1.6) with the change of enthalpy dH/dt on the right hand side. When rewriting eq. (1.6), the laws of equilibrium thermodynamics, i.e., the second class of physical laws, have already been used. Further, the enthalpy H may be expressed in terms of the temperature T (for constant pressure):

$$dH = d(Mh) = c_p \, d(MT) \tag{1.7}$$

eventually leading to

$$\dot{Q} + \dot{W}_S - \dot{Q}_{LL} = \left(Mc_p \frac{dT}{dt}\right)_{\text{in the vessel}} \tag{1.8}$$

for constant mass of the contents of the vessel with $T(t = 0) = T_I$.

The question regarding the variation of temperature with time, however, can not yet be answered with this equation alone. Apart from the laws of conservation and equilibrium, one needs the rate equations, i.e., one needs statements on the dependency of the fluxes on the field variables, such as the temperature, the flow velocity, and the concentration. These laws are always formulated in such a way that the fluxes vanish when approaching equilibrium. In the simplest version, the fluxes are taken as linearly related to the departure of the state variables from their equilibrium values. Should the steam in the jacket be in thermal equilibrium with the fluid in the vessel, then its temperature T_V would have to be equal to the temperature T of the vessels contents and vice versa:

$$T_{V,\text{Equilibrium}} = T \tag{1.9}$$

or

$$T_{\text{Equilibrium}} = T_V \tag{1.10}$$

$$\dot{Q} = K(T_V - T_{V,\text{Equilbrium}}) \tag{1.11}$$

The factor of proportionality K is thereby usually subdivided into two or more factors:

$$\dot{Q} = kA(T_V - T) \tag{1.12}$$

A is the transfer surface area, in this case, the surface area of the wall of the vessel, which is equipped with the jacket. The area specific proportionality factor $k = K/A$ is called "overall heat transfer coefficient." Analogous to eq. (1.12) one can write the rate equations for the heat losses in eqs. (1.2) and (1.3):

$$\dot{Q}_{LJ} = (kA)_{LJ} (T_V - T_a) \tag{1.13}$$

$$\dot{Q}_{LL} = (kA)_{LL}(T - T_a) \tag{1.14}$$

Since the streams are vectors, one has to exercise due care that their direction is always the same in the balance and in the corresponding rate equation. If the power of the stirrer $\dot{W}_S$ is known, e.g., kept constant by an appropriate control, the question posed under (a) in section 1.2 can be answered from the combined application of equations (1.8), (1.12), and (1.14). The list of parameters in this case contains nine quantities:

$$\text{parameters} = (\dot{Q}, \dot{W}_S, \dot{Q}_{LL}, Mc_p, kA, (kA)_{LL}, T_V, T_I, T_a)$$

1.4 Development in Terms of the Unknown Quantity

The sought-after quantity $T(t)$, the temperature of the contents of the vessel, can now be calculated from the three equations (1.8), (1.12), and (1.14). The not-sought-after quantities $\dot{Q}$, $\dot{Q}_{LL}$, which depend on T, however, are eliminated from the three equations. That can simply be achieved by inserting eq. (1.12) and (1.14) into (1.8):

$$kA(T_V - T) + \dot{W}_S - (kA)_{LL}(T - T_a) = Mc_p\frac{dT}{dt} \tag{1.15}$$

$$T(t = 0) = T_I \tag{1.16}$$

This is a first-order ordinary differential equation for $T(t)$, that can be solved by separation of variables (T, t), if the remaining seven parameters $(kA, \dot{W}_S, (kA)_{LL}, Mc_p, T_V, T_a, T_I)$ are known data. Before the rigorous solution, it is desirable to reduce the number of variables and parameters by casting them into non-dimensional form. The parameters kA and Mc_p are easily combined with the variable t to form a non-dimensional time variable

$$\tau \equiv \frac{kAt}{Mc_p} \tag{1.17}$$

This means that time is no longer measured in seconds, minutes, or hours but in terms of the time $t_c = Mc_p/(kA)$ characteristic of the given problem. Temperature is replaced by a normalized temperature difference

$$\vartheta \equiv \frac{T_V - T}{T_V - T_I} \tag{1.18}$$

Then eq. (1.15) transforms to

$$\vartheta + \omega - \varkappa(\vartheta_a - \vartheta) = -\frac{d\vartheta}{d\tau} \qquad \vartheta(\tau = 0) = 1 \tag{1.19}$$

In that form, the equation has only three parameters in place of seven:

$$\omega \equiv \frac{\dot{W}_S}{kA(T_V - T_I)} \tag{1.20}$$

$$\varkappa \equiv \frac{(kA)_{LL}}{kA} \qquad (1.21)$$

$$\vartheta_a \equiv \frac{T_V - T_a}{T_V - T_1} \qquad (1.22)$$

By introducing

$$\tau^{\cdot} \equiv (1 + \varkappa)\tau \qquad (1.23)$$

and

$$\vartheta^{\cdot} = \vartheta + \frac{\omega - \varkappa\,\vartheta_a}{1 + \varkappa} \qquad (1.24)$$

the equation can be further condensed:

$$\vartheta^{\cdot} = -\frac{d\vartheta^{\cdot}}{d\tau^{\cdot}} \qquad (1.25)$$

$$\vartheta^{\cdot}(\tau^{\cdot} = 0) = 1 + \frac{\omega - \varkappa\,\vartheta_a}{1 + \varkappa} = \vartheta_I^*$$

In this extremely compact form, it contains only one parameter of ϑ_I^*, which also might be avoided by choosing the variable $\vartheta^*/\vartheta_I^*$.

1.5 Mathematical Solution

The dimensional equation (1.15) with its seven parameters can, no doubt, be solved mathematically. The solution becomes algebraically simpler however, if one has taken the pains to bring it to the most compact form of eq. (1.25) with only one parameter:

$$-\int_{\vartheta_I^*} \frac{d\vartheta^{\cdot}}{\vartheta^{\cdot}} = \int_0 d\tau^{\cdot} \qquad (1.26)$$

$$\tau^* = -\ln \frac{\vartheta^*}{\vartheta_I^*} \qquad (1.27)$$

the solution being:

$$\vartheta^* = \vartheta_I^* \exp(-\tau^*) \qquad (1.28)$$

The solution has, therefore, the same simple form that is obtained when neglecting the power of the stirrer and the heat losses ($\omega = 0$, $\varkappa = 0$).

1.6 Discussion of the Results

The answer to the question (a) in section 1.2 regarding the temperature of the vessel contents as a function of time is given formally be eq. (1.28) together with eqs. (1.17) and (1.18), the dimensionless quantities being defined by eqs. (1.20) to (1.25). Now one has to check whether the result is in agreement with physical experience and especially whether all possible limiting cases are correctly described by the solution. With the values of the parameters suitably chosen, one can describe and possibly even extrapolate experimental data. The initial condition $\vartheta^*(\tau^* = 0) = \vartheta_I^*$ is, of course, fulfilled by eq. (1.28). The initial rate of change of temperature is

$$
\left(\frac{\mathrm{d}T}{\mathrm{d}t}\right)_{t=0} = (T_V - T_I)\frac{kA}{Mc}\,[1 + \omega + \varkappa(1 - \vartheta_a)] \tag{1.29}
$$

i.e., the heat loss has no influence on the initial slope of the temperature-time curve, if the initial temperature T_I is equal to the ambient temperature T_a (in this case $\vartheta_a = 1$). For longer times, the steady state temperature can be found from eq. (1.24) with $\vartheta_\infty^* = 0$ as

$$
\frac{T_V - T_\infty}{T_V - T_I} = \frac{\varkappa\vartheta_a - \omega}{1 + \varkappa} = \vartheta_\infty \tag{1.30}
$$

The steady state temperature T_∞ can be higher or lower than the vapor temperature depending on whether the power of the stirrer or the heat loss from the lid is higher. It may be surprising that the heat loss from the steam jacket to the ambient $\dot{Q}_{LJ}$, according to eq. (1.14), plays no role at all in the answer to the question (a) in section 1.2. A little reflection will easily lead to the reason thereof. From eq. (1.30), one can also recognize that ϑ^* of eq. (1.24) is just $\vartheta - \vartheta_\infty$, a measure of the approach to steady state.

By setting $\mathrm{d}T/\mathrm{d}t = 0$ in eq. (1.15), T_∞ and, therefore, the final compact form of the solution could be more easily determined as (exercise):

$$
\frac{T - T_\infty}{T_I - T_\infty} = \exp\left(-\frac{kA + (kA)_{LL}}{Mc_p}t\right) \tag{1.31}
$$

$$
T_\infty = T_V + \frac{\dot{W}_V - (kA)_{LL}(T_V - T_a)}{kA + (kA)_{LL}} \tag{1.32}
$$

In this form, the result can be most conveniently understood and discussed. The time constant of the heating process is

$$
t_c = \frac{Mc_p}{kA + (kA)_{LL}} = \frac{\varrho c_p V}{kA(1 + \varkappa)} \tag{1.33}
$$

Its order of magnitude can be easily estimated thus: For many liquids, the volumetric heat capacity is $(\varrho\, c_p)_l \approx 2\cdot 10^6$ J/(m^3 K) (for water, $4.2\cdot 10^6$ J/(m^3 K)). The overall heat transfer coefficient between the jacket steam and the vessel contents mainly depends on the convective coefficient on the inside of the vessel wall. As a rough estimate, one may use $k = 1000$ W/(m² K). If the temperature of the lid is not very different from that of the ambient, k_{LL} will be on the order of 1 to 10 W/(m² K) (free convection and radiation on the outside of the lid); and $(kA)_{LL}/(kA)$ will usually be small compared to unity. Assuming that the cylinder bottom is flat and is heated, the ratio of cylinder volume to active surface area for the vessel is

$$\frac{V}{A} = \frac{\pi D^2 L}{4(\pi DL + \pi D^2/4)} = \frac{D}{4 + D/L} \tag{1.34}$$

L is the heated height of the cylindrical shell ≈ height of liquid. From this, we find the time constant to be

$$t_c^* \approx 4 \cdot 10^3 \text{ s } \frac{D/m}{4 + D/L} \tag{1.35}$$

For aqueous liquids, it is around 800 seconds for a vessel diameter of 1 m and $D/L = 1$. The time required to reduce the temperature difference to 1% of the initial value is

$$t_{99\%} = t_c^* \ln 100 \tag{1.36}$$

and, with the above assumptions, it is roughly one hour.

The further questions (b) and (c) posed in section 1.2, in respect of the steam consumption and the influence of the stirrer speed on the heating process may also be answered now. The steam consumption is obtained by rearranging the energy balance for the steam jacket (eq. 1.2) as:

$$\dot{M}_V = \frac{\dot{Q} + \dot{Q}_{LJ}}{\Delta h_v} \tag{1.37}$$

$\dot{Q}$ and $\dot{Q}_{LJ}$ are calculated from eqs. (1.12) and (1.13). The maximum steam consumption occurs at the beginning of the heating process:

$$\dot{M}_{V,I} = \frac{kA(T_V - T_I) + (kA)_{LJ}(T_V - T_a)}{\Delta h_V} \tag{1.38}$$

Whether it is possible to provide the steam required for the initial heating and the pressure loss in the jacket for this high steam flow rate should also be checked.

The mechanical power dissipated and the overall heat transfer coefficient are both affected by the stirrer speed. The mechanical power is calculated as the product of the drag force $\dot{F}_D$ on the stirrer and the velocity ω. With $F_D = \rho \omega^2 d_s^2 c_D$ and $\omega = nd_s$

$$\dot{W}_S = \rho d_s^5 n^3 c_D \tag{1.39}$$

The drag coefficient so defined depends on a Reynolds number

$$Re = \frac{\omega d_S}{\nu} = \frac{n d_S^2}{\nu} \tag{1.40}$$

on the dimensions of the stirrer and the vessel, and, possibly, on other criteria such as the Froude number. For $Re > 10^4$, constant values of c_D ranging from 0.2 to 20 are reached. With $c_D \approx 2$, $d_s \approx 1$ m, $\rho = 10^3$ kg/m^3 and a rotational speed of $n = 1/s$, the calculated stirrer power is 2 kW. For water with $\nu = 10^{-6}$ m^2/s, the Reynolds number wold be around 10^6 in this case. For $k = 1000$ W/(m^2 K) and $A = 4$ m^2, the stirrer would increase the steady state temperature by

$$\Delta T_S \approx \frac{\dot{W}_S}{kA} \approx 0.5 \text{ K} \tag{1.41}$$

On the other hand, the heat loss over the lid, with $k_{LL} = 10$ W/(m^2 K), $A_{LL} = 1$ m^2 and $(T_V - T_a) = 100$ K, would decrease it by 0.25 K. In such cases, neglecting the heat loss over the lid and the stirrer power would cause little error. With increasing stirrer speed, the heat transfer coefficient on the inside of the vessel wall will also increase as $\alpha_i \propto n^{2/3}$ [H3, pp. 3.14.3, V1, pp. Ma1-8]. Since the other resistances due to the condensate film and the wall of the vessel are usually small compared to $1/\alpha_i$, the overall heat transfer coefficient will also increase approximately as $k \propto n^{2/3}$. The increase in the steady state temperature due to the stirrer power is thus related to speed as $n^{2.33}$ for higher speeds and roughly proportional to the square of the speed for lower speeds (for lower Reynolds numbers, $c_D \propto 1/Re$ and $\alpha_i = $ const.). At high Reynolds numbers, ΔT_s would be increased by a factor of five (see eq. [1.41]) if the stirrer speed is doubled. The heating time would be reduced by about 37% ($t_c \propto n^{-2/3}$). This, however, would require an eightfold power for the stirrer drive (see eq. [1.39]).

Problems

1.1 Equation (1.12) is valid for an instant t, when the contents of the vessel have reached the temperature T. To calculate the heat Q, which has been flowing through the vessel wall from the beginning of the heating process, one can write $Q = (kAt)\Delta T_M$. Calculate the appropriate mean temperature difference ΔT_M in terms of only the initial and the final temperature differences ($\Delta T_I = T_V - T_I$, $\Delta T_F = T_V - T_F$). Stirrer power and heat losses may be neglected.

Solution: $\Delta T_M = \dfrac{(\Delta T_I - \Delta T_F)}{\ln(\Delta T_I / \Delta T_F)}$

1.2 Answer question (a) in section 1.2 for the case that the maximum available mass rate of steam $\dot{M}_{V,max}$ is just half the initial steam consumption calculated from eq. (1.38) (stirrer power and heat loss negligible!)

Solution: $\vartheta = 1 - \tau/2$ (for $\tau \leq 1$); $\vartheta = (1/2)e^{-(\tau-1)}$ (for $\tau > 1$).

1.3 How does the liquid level in the vessel change with time, if the boiling temperature T_b of the liquid is less than $T_\infty (T_\infty \approx T_V, t = 0$ for $T = T_b)$?

Solution: $(L/L_1) = \exp(-t/t_c), \; t_c = D_{\rho l} \, \Delta h_v/[4 \, k \, (T_V - T_b)]$

1.4 Calculate the outlet temperature T'' of a fluid flowing steadily through a steam-heated stirred tank (mass rate $\dot{M}_{in} = \dot{M}_{out} = \dot{M}$, specific heat capacity c_p, inlet temperature T', kA, and T_V are given, stirrer power and heat losses are negligible. Outlet temperature = temperature of the contents in steady state).

Solution: $T'' = T' + (T_V - T')N/(1 - N); \; N = kA/(\dot{M}c_p)$.

2 DOUBLE-PIPE HEAT EXCHANGER IN PARALLEL AND COUNTERFLOW

2.1 Description

Double-pipe apparatuses (Fig. 1.3) are relatively simple to produce and are preferred for high pressure applications. The two streams, in the inner tube and in the annulus formed between the inner and outer tubes, can be directed in parallel or in counter-flow to each other as shown by arrows in Fig. 1.4. Furthermore a coordinate z in the direction of the tube side flow has been introduced.

2.2 Formulation of the Questions

a. With the mass flow rates $\dot{M}_1$, $\dot{M}_2$, the pressure drops Δp_1, Δp_2, and the inlet and outlet temperatures T_1', T_1'', T_2', T_2'' given, the size of a heat exchanger is to be determined (surface area A, flow cross sections S_1, S_2) (design problem).
b. For a given apparatus, given fluids, mass flow rates, and inlet temperatures, the outlet temperatures and, if necessary, other quantities of interest such as the heat

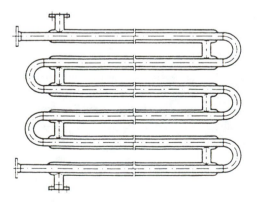

Figure 1.3 The double-pipe heat exchanger.

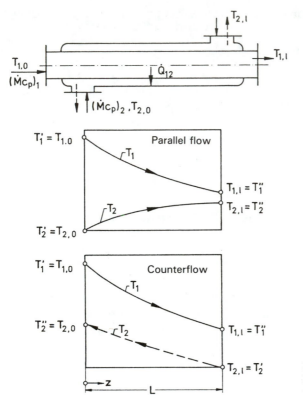

Figure 1.4 Sketch of double-pipe heat exchanger with temperature variation in parallel and counterflow.

transferred, pressure drop, and pumping power are to be calculated (rating problem).

2.3 Application of Physical Laws

When the sum of the kinetic and potential energies at the inlet and outlet are equal, the steady flow energy balance applied to the apparatus is

$$\dot{M}_1(h_1' - h_1'') + \dot{M}_2(h_2' - h_2'') - \dot{Q}_L = 0 \qquad (1.42)$$

A reasonable assumption in heat exchanger analysis is that the heat transfer between the apparatus and surroundings is small (due to small temperature differences between the outer stream and the surroundings or by good thermal insulation). Then, it follows from eq. (1.42) that the change in specific enthalpy of each stream is inversely proportional to its mass flow rate:

$$\frac{h_2' - h_2''}{h_1'' - h_1'} = \frac{\dot{M}_1}{\dot{M}_2} \qquad (1.43)$$

The ratio of the enthalpy changes above may be expressed in terms of the corresponding temperature changes alone. For pure substances one has

$$\varrho\, dh = \varrho c_p\, dT + (1 - \beta T)\, dp \tag{1.44}$$

Here the product of the thermal expansion coefficient β and the absolute temperature T is equal to unity for an ideal gas, i.e., the enthalpy is a function only of temperature. For liquid water at $20\,°C$ $\beta T \approx 0.06$ and $\rho = 10^3\ kg/m^3$. The strong dependency of enthalpy on temperature $(\partial h/\partial T)_p = c_p \approx 4.2 \cdot 10^3\ J/(kg\ K)$ is dominant relative to the weak dependency on pressure $(\partial h/\partial p)_T = (1 - \beta T)/\rho \approx 0.94 \cdot 10^{-3}\ J/(kg\ Pa)$. Even for a pressure change as large as 10^5 Pa, the influence on enthalpy is less than 2.3% of the influence of a temperature change of only 1K! In these cases, one can also write in place of eq. (1.43):

$$\frac{T_{2,1} - T_{2,0}}{T_{1,0} - T_{1,1}} = \frac{(\dot{M} c_p)_1}{(\dot{M} c_p)_2} \tag{1.45}$$

The energy balances for the inner tube (fluid 1) and the annulus (fluid 2) are, respectively,

$$(\dot{M} c_p)_1 (T_{1,0} - T_{1,1}) - \dot{Q}_{12} = 0 \tag{1.46}$$

$$(\dot{M} c_p)_2 (T_{2,0} - T_{2,1}) - \dot{Q}_{21} = 0 \tag{1.47}$$

and, thus, using eq. (1.45), it follows that

$$\dot{Q}_{21} = -\dot{Q}_{12} \tag{1.48}$$

For the heat transferred from fluid 1 to fluid 2, one can write the rate equation

$$\dot{Q}_{12} = kA(T_1 - T_2) \tag{1.49}$$

The temperatures T_1 and T_2, however, are not constant over the surface area A (or the coordinate z), so that an appropriate mean value of the temperature difference $(T_1 - T_2)_M$ has to be used in eq. (1.49). To determine this mean value, the variation of the temperature difference with z must be known. Consequently, we first have to apply the balances of eq. (1.46) and (1.47) and the rate equation (1.49) only locally, to the control volumes $dV_1 = S_1 dz$ and $dV_2 = S_2 dz$:

$$-(\dot{M} c_p)_1\, dT_1 - d\dot{Q}_{12} = 0 \tag{1.50}$$

$$-(\dot{M} c_p)_2\, dT_2 - d\dot{Q}_{21} = 0 \tag{1.51}$$

$$d\dot{Q}_{12} = k(T_1 - T_2)\, dA \tag{1.52}$$

2.4 Development in Terms of the Unknown Quantities

By eliminating the heat rate $d\dot{Q}_{12}$ from the balance equations

$$-dT_1 = (T_1 - T_2)\frac{k\,dA}{(\dot{M}c_p)_1} \tag{1.53}$$

$$dT_2 = (T_1 - T_2)\frac{k\,dA}{(\dot{M}c_p)_2} \tag{1.54}$$

one obtains a system of two coupled ordinary differential equations for the temperatures T_1 and T_2 as functions of z: $dA = A\,dz/L$. With the "number of transfer units" (see also τ according to eq. [1.17]) defined by

$$N_i \equiv \frac{kA}{(\dot{M}c_p)_i} \qquad (i = 1, 2) \tag{1.55}$$

and the normalized variables for length

$$Z = \frac{z}{L} \tag{1.56}$$

and temperatures

$$\vartheta_1 \equiv \frac{T_1 - T_2'}{T_1' - T_2'} \tag{1.57}$$

$$\vartheta_2 \equiv \frac{T_2 - T_2'}{T_1' - T_2'} \tag{1.58}$$

the differential eqs. (1.53) to (1.54) become

$$\boxed{-\frac{d\vartheta_1}{dZ} = N_1(\vartheta_1 - \vartheta_2)} \tag{1.59}$$

$$\boxed{\frac{d\vartheta_2}{dZ} = N_2(\vartheta_1 - \vartheta_2)} \tag{1.60}$$

Adding these two equations, one obtains a single one for the temperature difference:

$$-\frac{d(\vartheta_1 - \vartheta_2)}{dZ} = (N_1 + N_2)(\vartheta_1 - \vartheta_2) \tag{1.61}$$

The normalization of the temperatures can be chosen arbitrarily and the form chosen here in eqs. (1.57) and (1.58) sets the entrance temperatures of the two streams to the convenient values $\vartheta_1' = 1$ and $\vartheta_2' = 0$.

2.5 Mathematical Solution

Equation (1.61) can be solved straightaway by separation of variables and integration:

$$-\ln \frac{(\vartheta_1 - \vartheta_2)}{(\vartheta_1 - \vartheta_2)_0} = (N_1 + N_2)Z \tag{1.62}$$

Integrating up to $Z = 1$ ($z = L$) gives

$$\ln \frac{(\vartheta_1 - \vartheta_2)_0}{(\vartheta_1 - \vartheta_2)_1} = (N_1 + N_2) \tag{1.63}$$

Dividing the balance eqs. (1.46) and (1.47) by (kA), the sum ($N_1 + N_2$) can be expressed as:

$$(N_1 + N_2) = kA \frac{(T_1 - T_2)_0 - (T_1 - T_2)_1}{\dot{Q}_{12}} \tag{1.64}$$

so that the heat rate can be given in terms of kA and the four temperatures alone:

$$\boxed{\dot{Q}_{12} = kA \frac{(T_1 - T_2)_0 - (T_1 - T_2)_1}{\ln[(T_1 - T_2)_0/(T_1 - T_2)_1]} = kA\Delta T_{LM}} \tag{1.65}$$

Thereby, the appropriate mean value of the temperature difference required in eq. (1.49) is found. It is the logarithmic mean of the temperature differences at the positions $z = 0$ and $z = L$ (see also Problem 1.1 for comparison). Inserting the exponential form of eq. (1.62) into eq. (1.59) and integrating again leads to the solution for $\vartheta_1(Z)$:

$$\frac{\vartheta_{1.0} - \vartheta_1(Z)}{\vartheta_{1.0} - \vartheta_{2.0}} = \frac{1 - e^{-(N_1+N_2)Z}}{1 + N_2/N_1} \tag{1.66}$$

By the same procedure, using eq. (1.60), $\vartheta_2(Z)$ can be obtained.

2.6 Discussion of the Results

To answer the question 2.2 (a), eq. (1.46) or (1.47), together with eq. (1.65), can be used to calculate the surface area required:

$$\dot{Q}_{12} = (\dot{M}c_p)_1 (T_1' - T_1'') = |\dot{M}c_p|_2 (T_2'' - T_2')$$

$$A_{req} = \frac{\dot{Q}_{12}}{k\Delta T_{LM}}$$

The overall heat transfer coefficient k depends on the flow velocities which are determined by the flow cross sections S_1 and S_2 chosen and the pressure drops for fluids 1 and 2. The outlet temperatures (question 2.2 [b]) can be found from eq. (1.66) with $Z = 1$. In the case of parallel flow, this can be done directly with $\vartheta_{1.0} = \vartheta_1' = 1$ and

$\vartheta_{2,0} = \vartheta_2' = 0$:

$$T_1'' = T_1' + \epsilon_1(T_1' - T_2')$$

with

$$\varepsilon_1 = \frac{1 - e^{-(N_1 + N_2)}}{1 + N_2/N_1} \tag{1.67}$$

For counterflow with $\vec{N}_2 < 0$ (z-coordinate against the flow direction of $\dot{M}_2$), $\vartheta_{2,1} = 0$, while $\vartheta_{2,0} = \vartheta_2''$ is the yet-unknown outlet temperature of stream 2. This can be expressed through the overall balance eq. (1.45) in terms of ϑ_1'' and the capacity flow rate ratio,

$$C \equiv \frac{(\dot{M}c_p)_1}{(\dot{M}c_p)_2} = \frac{N_2}{N_1} = \frac{\vartheta_{2,1} - \vartheta_{2,0}}{1 - \vartheta_1''} \tag{1.68}$$

leading to

$$\varepsilon_1 = \frac{1 - e^{-(1+C)N_1}}{1 + C\,e^{-(1+C)N_1}} \tag{1.69}$$

In the special case $C = -1$, i.e., counterflow with equal flow capacities ($N_1 - N_2 = 0$), eq. (1.69) leads to an indeterminate expression. By series expansion of the exponential function, one obtains

$$\varepsilon_1 = \frac{N_1}{1 + N_1} \qquad (C = -1) \tag{1.70}$$

The exponential function degenerates in this case to a linear function as may be seen from eq. (1.61). The temperature difference remains the same at any position. The quantity ϵ_i, introduced in eq. (1.67), is a dimensionless change of temperature and is usually called heat exchanger effectiveness, or efficiency.

$$\varepsilon_1 = 1 - \vartheta_1'' \tag{1.71}$$

$$\varepsilon_2 = \vartheta_2'' - 0 \tag{1.72}$$

It is generally defined as

$$\varepsilon_i \equiv \frac{(\text{change of temperature of stream } i)}{(\text{max. temperature difference})}$$

$$\varepsilon_1 \equiv \frac{T_1' - T_1''}{T_1' - T_2'} \tag{1.73}$$

$$\varepsilon_2 \equiv \frac{T_2'' - T_2'}{T_1' - T_2'} \tag{1.74}$$

From eq. (1.68) with (1.71) and (1.72) or directly from eq. (1.45), we see that the ratio $\varepsilon_2/\varepsilon_1$ is related to NTU and C as

$$\frac{\varepsilon_2}{\varepsilon_1} = \left|\frac{(\dot{M}c_p)_1}{(\dot{M}c_p)_2}\right| = \left|\frac{N_2}{N_1}\right| = |C| = R \tag{1.75}$$

Through the balances of eq. (1.46) and (1.47), the effectiveness can also be written as a dimensionless heat rate:

$$\boxed{\frac{\dot{Q}_{12}}{\dot{M}_i c_{pi}(T_1' - T_2')} = \frac{kA}{\dot{M}_i c_{pi}} \cdot \frac{\Delta T_M}{T_1' - T_2'}} \tag{1.76}$$

$$\boxed{\varepsilon_i = N_i \cdot \Theta} \qquad (i = 1, 2)$$

which expresses the basic relationship between the three dimensionless quantities ε_i, N_i and Θ. If Θ is known in terms of ε_1, ε_2 then N_1, N_2 can be calculated (design problem). If Θ is known as a function of N_1, N_2, the values of ε_1, ε_2 can be found (rating problem!). Knowledge of the functions

$$\Theta(\varepsilon_1, \varepsilon_2) \tag{1.77}$$

$$\Theta(N_1, N_2) \tag{1.78}$$

is thus the key to the questions 2.2 (a), (b). The quantity Θ, as can be seen from eq. (1.76), is the integral mean temperature difference ΔT_M divided by the maximum temperature difference. For the cases of parallel flow and counterflow treated here, the mean temperature difference is the logarithmic mean, ΔT_{LM} (see eq. [1.65]).

Problems

1.5 Obtain the dimensionless mean temperature difference Θ in terms of N_1, N_2 for the stirred tank with steady streams from problem 1.4 (subscript "1" contents of tank, "2" steam)

Solution: $\Theta = 1/(1 + N_1)$, $N_2 = 0$, $C = 0$

1.6 Determine $\Theta(N_1)$ for $C = 0$ $[(\dot{M}c_p)_2 \to \infty]$ for parallel and counterflow.

Solution: $\Theta = (1 - \exp(-N_1))/N_1$

1.7 The following inlet and outlet temperatures have been measured in a heat exchanger operated in parallel flow: $T_1' = 100\,°C$, $T_2' = 20\,°C$, $T_1'' = 60\,°C$, $T_2'' =$

50 °C. What outlet temperatures T_{1c}'', T_{2c}'' would you expect to find under the same inlet conditions if operated in counterflow?

Solution: $T_{1c}'' = 53.6\,°C$, $T_{2c}'' = 54.8\,°C$ $(C = -3/4,\ N_1 = 1.188)$

1.8 Derive the function $\Theta\ (\epsilon_1,\ \epsilon_2)$ for parallel and counterflow heat exchangers.

Solution: Parallel flow
$$\Theta = -\frac{\varepsilon_1 + \varepsilon_2}{\ln[1 - (\varepsilon_1 + \varepsilon_2)]}$$

Counterflow
$$\Theta = \frac{\varepsilon_1 - \varepsilon_2}{\ln[(1 - \varepsilon_2)/(1 - \varepsilon_1)]} \qquad \varepsilon_1 \neq \varepsilon_2$$

$$\Theta = 1 - \varepsilon_1 = 1 - \varepsilon_2, \qquad \varepsilon_1 = \varepsilon_2$$

3 DOUBLE-PIPE BAYONET HEAT EXCHANGER

3.1 Description

Double-pipes with one end closed, as shown in Fig. 1.5, are often used as bayonet heating (or cooling) elements in various types of apparatus. They are also called Field tubes. The outer medium heated or cooled by such a Field tube (or double-pipe) will often be well mixed (as in stirred tanks, fluidized beds, etc.) or have a constant temperature T_b, as in a boiling liquid. The heating (or cooling) medium may be a liquid or gaseous heat carrier with mass flow rate M and specific heat capacity c_p (assumed constant). The medium may first enter the inner tube (case I) or the annulus (case II).

3.2 Formulation of the Questions

The earlier analysis of the double-pipe apparatus in parallel and counterflow might lead us to expect that the heat transferred from the heating medium to the well-mixed contents of the outer vessel (boiling liquid, fluidized bed, etc.) could again be calculated from a simple equation:

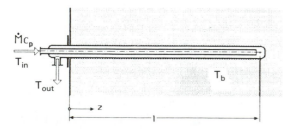

Figure 1.5 The double-pipe bayonet heat exchanger.

$$\dot{Q} = kA\Delta T_{\text{LM}} \qquad (1.79)$$

In the following sections, we shall check whether it is really that simple or if it is necessary to use a mean temperature difference other than the logarithmic mean. To accomplish this, the variation of the temperature of the heating medium in the inner tube and in the annulus has to be calculated as a function of the coordinate z.

3.3 Application of Physical Laws

As in the preceding sections, we shall combine the rate equations and the energy balance for each of the control volumes dV_i and dV_a to formulate the unknown temperature as a function of the length coordinate. With the abbreviations

$$\vartheta \equiv \frac{T - T_b}{T_{\text{in}} - T_b}; \qquad \frac{k_i A_i}{\dot{M}c_p} \equiv N$$

$$\zeta \equiv N\frac{z}{l}; \qquad \frac{k_a A_a}{k_i A_i} \equiv \varkappa \qquad (1.80)$$

the two differential equations are

$$-\frac{d\vartheta_i}{d\zeta} = \vartheta_i - \vartheta_a \qquad (1.81)$$

$$-\frac{d\vartheta_a}{d\zeta} = \vartheta_i - (1+\varkappa)\vartheta_a \qquad (1.82)$$

subject to the boundary conditions:

$$\vartheta_i(0) = 1 \qquad (1.83)$$

$$\vartheta_i(N) = \vartheta_a(N) \qquad (1.84)$$

Here it is formulated for the fluid entering the bayonet assembly by the inner tube (case I). For $\varkappa = 0$ and $N_2 = -N_1$, these differential equations are the same as for a counterflow heat exchanger (eqs. [1.59] and [1.60]). With the boundary condition $\vartheta_i(N) = \vartheta_a(N)$ however, $\varkappa = 0$ would be a trivial case and the temperature at the inlet, $\vartheta = 1$, would be maintained throughout.

The mathematical solution can be found by decoupling the two equations, i.e., by eliminating one of the two dependent variables ϑ_i, ϑ_a. Denoting the derivative with respect to ζ by a prime, $d\vartheta/d\zeta = \vartheta'$, we get from eq. (1.81)

$$\vartheta_a = \vartheta_i + \vartheta_i' \qquad (1.85)$$

This equation and its derivative

$$\vartheta'_a = \vartheta'_i + \vartheta''_i \tag{1.86}$$

can now be inserted into eq. (1.82). An ordinary differential equation of second order for $\vartheta_i(\zeta)$ results:

$$\vartheta''_i - \varkappa \vartheta'_i - \varkappa \vartheta_i = 0 \tag{1.87}$$

For $\vartheta_a(\zeta)$, a similar procedure (exercise) will yield

$$\vartheta''_a - \varkappa \vartheta'_a - \varkappa \vartheta_a = 0 \tag{1.88}$$

The general solutions to these two equations are

$$\vartheta_i(\zeta) = A \, e^{m_1 \zeta} + B \, e^{m_2 \zeta} \tag{1.89}$$

and

$$\vartheta_a(\zeta) = C \, e^{m_1 \zeta} + D \, e^{m_2 \zeta} \tag{1.90}$$

where m_1 and m_2 denote the roots of the characteristic eq. $m^2 - \varkappa m - \varkappa = 0$:

$$m_{1,2} = \left(1 \pm \sqrt{1 + \frac{4}{\varkappa}} \right) \frac{\varkappa}{2} \tag{1.91}$$

To determine the four constants, we have the two boundary conditions (1.83) and (1.84) and two other conditions derived from these with differential eqs. (1.81) and (1.82):

$$\vartheta'_i(N) = 0 \tag{1.92}$$

$$\vartheta'_a(N) = \varkappa \vartheta_a(N) \tag{1.93}$$

The somewhat laborious calculation of the four constants eventually leads to the solution for case I:

$$\vartheta_{i\,(I)}(\zeta) = \frac{m_1 \, e^{m_1 N} \, e^{m_2 \zeta} - m_2 \, e^{m_2 N} \, e^{m_1 \zeta}}{M} \tag{1.94}$$

$$\vartheta_{a\,(I)}(\zeta) = \frac{m_1 \, e^{m_2 N} \, e^{m_1 \zeta} - m_2 \, e^{m_1 N} \, e^{m_2 \zeta}}{M} \tag{1.95}$$

where

$$M = m_1 \, e^{m_1 N} - m_2 \, e^{m_2 N}$$

The temperature at the closed end of the bayonet is found (with $m_1 + m_2 = \varkappa$ and $\zeta = N$) to be:

$$\vartheta_{1\,(I)} = \vartheta_i(N) = \vartheta_a(N) = \frac{(m_1 - m_2)\,e^{\varkappa N}}{M} \tag{1.96}$$

The outlet temperature of the heating medium becomes

$$\vartheta_{\text{out (I)}} = \vartheta_a(0) = \frac{m_1\,e^{m_2 N} - m_2\,e^{m_1 N}}{M} \tag{1.97}$$

The solution for case II (fluid entering the bayonet by way of the annulus) can be found by replacing ζ by $-\zeta$ in the differential equations. In place of eqs. (1.87) and (1.88), we get

$$\vartheta_i'' + \varkappa\,\vartheta_i' - \varkappa\,\vartheta_i = 0 \tag{1.98}$$

and

$$\vartheta_a'' + \varkappa\,\vartheta_a' - \varkappa\,\vartheta_a = 0 \tag{1.99}$$

with the boundary conditions

$$\vartheta_a(0) = 1 \qquad\qquad \vartheta_a(N) = \vartheta_i(N) \tag{1.100}$$

$$\vartheta_i'(N) = 0 \qquad\qquad \vartheta_a'(N) = -\varkappa\,\vartheta_a(N) \tag{1.101}$$

After appropriate calculations (exercise!), we find that the solutions for case II can be expressed by

$$\vartheta_{a\,(II)}(\zeta) = e^{-\varkappa\zeta}\,\vartheta_{i\,(I)}(\zeta) \tag{1.102}$$

$$\vartheta_{i\,(II)}(\zeta) = e^{-\varkappa\zeta}\,\vartheta_{a\,(I)}(\zeta) \tag{1.103}$$

Therefore, it follows that the normalized temperature at the closed end in case II is lower than in case I by a factor of $e^{-\varkappa N}$:

$$\vartheta_{1\,(II)} = \vartheta_{a\,(II)}(N) = \vartheta_{i\,(II)}(N) = e^{-\varkappa N}\,\vartheta_{1\,(I)} \tag{1.104}$$

Nevertheless, the outlet temperature of the heating medium turns out to be the same in both cases:

$$\vartheta_{\text{out (II)}} = \vartheta_{i\,(II)}(0) = \vartheta_{a\,(I)}(0) = \vartheta_{\text{out (I)}} \tag{1.105}$$

3.4 Discussion of the Results

Figure 1.6 shows the variation of temperature in the double-pipe bayonet heat exchanger with $\varkappa = 3$ and $N = 1$ for the two cases. The heating medium will be reheated in the second pass after passing a minimum which is always at $z = 1$ in case II. Despite different temperature variations inside, the efficiency ϵ, defined as before (see eq. [1.73], but with the temperatures conveniently written with subscripts "in" and "out" in place of the primes and double primes which have been used here to

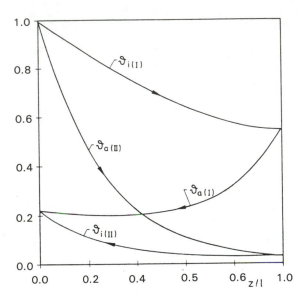

Figure 1.6 Temperature variation in the double-pipe bajonet heat exchanger.

denote the derivatives) is always the same for cases I and II ($\epsilon = 1 - \vartheta_{out}$). It depends on the two parameters $\varkappa$ and N. After some algebraic manipulation, it can be written as

$$\varepsilon(N_a, \varkappa) = 2\frac{1 - e^{-\mu N_a}}{(\mu + 1) + (\mu - 1)\,e^{-\mu N_a}} \tag{1.106}$$

with

$$\mu = \sqrt{1 + \frac{4}{\varkappa}}, \quad N_a \equiv \varkappa N, \quad \varkappa = \frac{k_a A_a}{k_i A_i} \quad N_a = \frac{k_a A_a}{M c_p}$$

Here the product $\varkappa N$ was chosen as a more relevant measure of the residence time of the heating medium (or the length coordinate). For any finite value of $\varkappa$, there is a maximum effectiveness

$$\varepsilon_{max}(\varkappa) = \lim_{N_a \to \infty} \varepsilon(N_a, \varkappa) = \frac{2}{\mu + 1} \tag{1.107}$$

For $\varkappa \to \infty$ and, consequently, $\mu \to 1$, i.e., for a perfectly insulated inner tube, the maximum possible effectiveness is obtained. In that case, one would have a constant temperature in the inner tube and the temperature curves for the annulus for both cases would be mirror images about a vertical axis through $Z = 0.5$. Equation (1.106) can be solved for N_a, so that, with $\Theta = \epsilon/N_a$, the mean temperature difference can also be given explicitly in terms of ϵ and $\varkappa$:

$$\Theta(\varepsilon, \varkappa) = \frac{\varepsilon\mu}{\ln\{[2 + (\mu - 1)\varepsilon]/[2 - (\mu + 1)\varepsilon]\}} \tag{1.108}$$

Now it is possible to answer the question raised in section 3.2, viz., whether the logarithmic mean temperature difference is the correct basis for calculating the heat transferred, as done in eq. (1.79). The logarithmic mean is easily calculated for constant temperature of the outer medium (i.e., $C = 0$):

$$\Theta_{LM}(\varepsilon) = -\frac{1 - (1 - \varepsilon)}{\ln(1 - \varepsilon)} \qquad (1.109)$$

Comparing this result with the mean temperature difference (eq. [1.108]), we see that eq. (1.79) would hold good only for $\varkappa \to \infty$, i.e., $\mu = 1$. However, in order to retain the logarithmic mean as the basis for calculations, one may introduce a LMTD correction factor F

$$F \equiv \frac{\triangle T_M}{\triangle T_{LM}} = \frac{\Theta}{\Theta_{LM}} \qquad (1.110)$$

The factor F describes the diminution of the performance compared to the ideal (counterflow) case:

$$F = \frac{\mu \ln(1 - \varepsilon)}{\ln\{[2 - (\mu + 1)\varepsilon]/[2 + (\mu - 1)\varepsilon]\}} \qquad (1.111)$$

Figure 1.7 shows the mean temperature difference Θ according to eq. (1.108) as a function of ϵ, with $\varkappa$ as a parameter. The straight lines through the origin are lines of constant NTU, $N_a = $ const. The double-pipe bayonet heat exchanger is obviously very unsuitable if the overall heat transfer coefficient at the outer wall becomes smaller than that at the inner wall. It may be recalled that the temperature of the

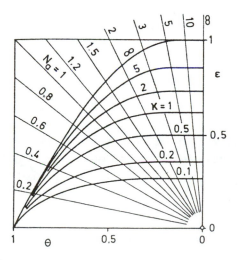

Figure 1.7 Mean temperature difference Θ as a function of efficiency ϵ with $\varkappa = (kA)_0/(kA)_i$ as parameter.

medium outside the double pipe (T_b) was assumed to be constant for this analysis. This is possible either for very large heat capacity of the outer medium ($C \rightarrow 0$, for example, with evaporation or condensation) or for ideally mixed flow of the stream $\dot{M}_2$. In the latter case, however, T_b would not be the inlet temperature but the outlet temperature of medium 2. In this case, ϵ is defined with $T_b = T_2''$ in place of T_2'. Under the stated assumptions, an effectiveness ϵ_1 for the medium 1 inside the double-pipe, according to the usual definition (see eq. [1.73])

$$\varepsilon_1 = \frac{T_1' - T_1''}{T_1' - T_2'} \tag{1.112}$$

is related to ϵ and C (exercise!) as

$$\varepsilon_1 = \frac{\varepsilon}{1 + C\varepsilon} \tag{1.113}$$

This, of course, gives the original definition again for $C \rightarrow 0$.

3.5 Pressure Drop in Double-Pipes

In order to force liquids or gases through the pipes and annuli of double-pipe apparatuses, pumping power is required, which is proportional to the pressure drop for given flow rates.

For the design of heat exchangers, we also need, apart from the laws of heat transfer (to calculate the overall heat transfer coefficients), a knowledge of fluid mechanics. Here, primarily, the relationships between pressure drop, flow rate, viscosity and density of the fluid, and the geometry of the duct are of importance.

For fully developed *laminar flow* of incompressible Newtonian fluids in a circular duct of diameter D or in an annulus whose outer and inner passage dimensions are, respectively, D and $K \cdot D$ (i.e., the inner diameter of the outer tube and the outer diameter of the inner tube forming the annulus), the Hagen-Poiseuille law in a form generalized for annuli holds [B5]:

$$\left[1 - K^4 + \frac{(1 - K^2)^2}{\ln K} \right] \triangle p = \frac{128 \eta \dot{M} L}{\pi \varrho D^4} \tag{1.114}$$

For a circular duct, K is equal to zero and the expression in K within brackets becomes unity. For the general case, the average velocity can be written as

$$u = \frac{4 \dot{M}}{\pi \varrho D^2 (1 - K^2)} \tag{1.115}$$

Utilizing the following definitions for the hydraulic diameter and Reynolds number

$$d_h = 4 \frac{\text{cross sectional area}}{\text{wetted perimeter}} = D(1 - K) \qquad (1.116)$$

and

$$Re = \frac{\rho u d_h}{\eta} = \frac{4 \dot{M}}{\pi D(1 + K)\eta} \qquad (1.117)$$

we can write

$$\frac{\Delta p}{\rho u^2/2} = \frac{64\varphi(K)}{Re} \frac{L}{D_h} \qquad (1.118)$$

The function $\varphi(K)$ is given by:

$$\varphi(K) = \frac{(1 - K)^2}{1 + K^2 + (1 - K^2)/\ln K} \qquad (1.119)$$

For circular ducts ($K = 0$), the function $\varphi(0)$ has the value 1. $\varphi(K)$ shows a steep initial gradient and reaches a limiting value of 3/2 for $K = 1$ (parallel plates duct). Figure 1.8 shows the graph of the function $\varphi(K)$.

For a bayonet (i.e., a double-pipe arrangement closed at one end) of length L, outer and inner dimensions of the annulus D and KD and the ratio of the pipe wall thickness to diameter δ, the internal diameter of the core tube is $(1 - 2\delta)KD$. Not considering turnaround losses, the expression for the frictional pressure drop through the bayonet following eq. (1.114) becomes

$$\frac{\Delta p}{\Delta p_0} = [(1 - 2\delta)K]^{-4} + \left(1 - K^4 + \frac{(1 - K^2)^2}{\ln K}\right)^{-1} \qquad (1.120)$$

Here Δp_0 is the pressure drop in a simple circular duct of diameter D and is obtained by setting $K = 0$ in eq. (1.114). Equation (1.120) is plotted in Fig. 1.9 as a function of the diameter ratio K for $\delta = 0$ (negligible wall thickness) and $\delta = 0.2$. For small inner tube diameters, the main flow resistance lies in the inner tube and decreases as $[(1 - 2\delta)K]^{-4}$. For larger K, the resistance is essentially in the annulus and increases

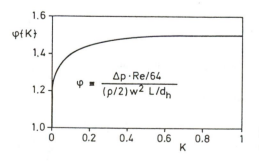

Figure 1.8 Factor $\varphi(K)$ for frictional pressure drop in an annulus.

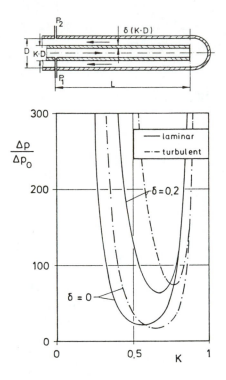

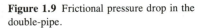

Figure 1.9 Frictional pressure drop in the double-pipe.

correspondingly steeply with decreasing gap width. As can be seen in the figure, a well-marked minimum in pressure drop occurs at a certain diameter ratio. For turbulent flow, the valid equation in place of eq. (1.118) is

$$\frac{\triangle p}{(\varrho/2)u^2} = \xi(Re)\frac{L}{d_\mathrm{h}} \tag{1.121}$$

In fully *turbulent flow*, $\xi(Re)$ is a weak function for smooth pipes and a constant for rough pipes [H3, V1]. Contrary to laminar flow [$\xi = 64\varphi(K)/Re$], ξ is practically independent of the cross-sectional shape of the duct in turbulent flow. So, for turbulent flow in a circular duct as well as in an annulus, we obtain, to sufficient approximation,

$$\boxed{(1 - K)^3(1 + K)^2\triangle p = \xi\frac{8\dot{M}^2 L}{\pi^2\varrho D^5}} \tag{1.122}$$

The dependency of pressure drop on diameter in turbulent flow is as D^{-5} and is stronger than for laminar flow ($\propto D^{-4}$, eq. [1.114]). Similar to eq. (1.120) for laminar flow in a double-pipe bayonet, we obtain for turbulent flow (with $\xi = $ constant)

$$\frac{\triangle p}{\triangle p_0} = [(1 - 2\delta)K]^{-5} + [(1 - K)^3(1 + K)^2]^{-1} \tag{1.123}$$

The values calculated from this equation for $\delta = 0$ and $\delta = 0.2$ are shown as chain lines on Fig. 1.9. To this, the flow turnaround losses have to be added, especially for shorter tubes.

For turbulent flow in smooth tubes, the friction factor ξ can be calculated, for example, from using an equation due to Filonenko [H3, V1]:

$$\boxed{\xi = (1,82 \lg Re - 1,64)^{-2}} \qquad (1.124)$$

Using this equation, the relative pressure drop can no longer be given in terms of K and δ alone. This can be achieved, however, by using the simpler power law as given by Blasius

$$\xi = \frac{0.316}{Re^{1/4}} \qquad (1.125)$$

or any similar power law expression:

$$\xi = \frac{\text{const}}{Re^m} \qquad (1.126)$$

Using eq. (1.126), we arrive at the relative pressure drop:

$$\frac{\Delta p}{\Delta p_0} = [(1 - 2\delta)K]^{-(5-m)} + [(1 - K)^3(1 - K)^{2-m}]^{-1} \qquad (1.127)$$

When $m = 0$, we obtain eq. (1.123). With $m = 1/4$, the calculated value differs little from that for $m = 0$. Thus, an additional curve for $m = 1/4$ has not been shown in Fig. 1.9.

3.6 Heat Transfer in Double-Pipes

The heat transfer coefficients for flow in tubes and annuli can be calculated for simple boundary conditions, e.g., constant temperature or constant heat flux at the wall, for which well-known formulas are available in textbooks and, in particular, in handbooks on heat transfer [H3, V1]. For fully developed laminar flow and sufficiently long ducts, asymptotic values of the Nusselt numbers are obtained, e.g.,

$$\boxed{Nu_{T\infty} \approx 3.66 + 1.2K^m} \qquad (1.128)$$

where $m = -0.8$ for outer walls. If K is set equal to zero or unity, respectively, one obtains the corresponding values for circular duct and parallel plates (one side adiabatic). No standard formula can be found, however, in the handbooks [H3, V1] for the case of a double-pipe bayonet heat exchanger with heat flow on both sides of the annulus at different temperatures. The equation given in these sources for heat flow on both sides of the annulus is only valid for the same temperature on both walls. In the case that the two walls are at different temperatures, the asymptotic Nusselt numbers can be calculated from the known steady state temperature and velocity profiles. The steady state temperature profile in the annulus is

$$\vartheta = \frac{\ln x}{\ln K} \tag{1.129}$$

and the velocity profile [B5]

$$\frac{u}{\bar{u}} = 2\frac{1 - x^2 + 2x_m^2 \ln x}{1 + K^2 - 2x_m^2} \tag{1.130}$$

Here x is a normalized radius: $x = r/R$, whose value lies in the range $K \leq x \leq 1$, and x_m is the coordinate of the maximum velocity:

$$2x_m^2 = \frac{1 - K^2}{\ln(1/K)} \tag{1.131}$$

With the dimensionless heat flux at the outer wall of the annulus ($x = 1$)

$$\Phi_a = \frac{-2(1 - K)}{\ln K} \tag{1.132}$$

and at the inner wall ($x = K$)

$$\Phi_i = \frac{\Phi_a}{K} \tag{1.133}$$

and the caloric average temperature

$$\bar{\vartheta} = \frac{2}{1 - K^2} \int_K^1 \frac{u}{\bar{u}} (x) \, \vartheta(x) \, x \, dx \tag{1.134}$$

the Nusselt numbers based on the hydraulic diameter as a characteristic length (according to eq. [1.117]) now become

$$Nu_a = \frac{\Phi_a}{\bar{\vartheta}} \tag{1.135}$$

and

$$Nu_i = \frac{\Phi_i}{1 - \bar{\vartheta}} \tag{1.136}$$

Equation (1.134), together with eqs. (1.129) and (1.130) for the temperature and velocity profiles, can be integrated in closed form (exercise!) to give

$$\bar{\vartheta} = \frac{(2x_m^2 - K^2)^2 + 2x_m^2(K^2 - 3)/4}{(1 - K^2)(2x_m^2 - K^2 - 1)} \qquad (1.137)$$

Nusselt numbers calculated from the equations (1.131) to (1.137) agree with those given by Shah and London [S8] in tabular and in graphical form. For $K \to 0$ (thin wire in the axis of a tube), the flux Φ_i and the Nusselt number Nu_i at the inner surface tend to infinity. However, the product $K\Phi_i = \Phi_a$ or (KNu_i), respectively, tend to zero as $1/\ln K$. Nu_a, in this limit, tends to a finite value

$$\lim_{K \to 0} Nu_a = \frac{8}{3} = 2.667 \qquad (1.138)$$

For $K \to 1$ (parallel plates, gap width $R(1 - K) \ll$ radius R), a limit analysis yields the value $1/2$ for ϑ and $\Phi_0 = \Phi_i = 2$. So, the Nusselt numbers, in this case, reach the value

$$\lim_{K \to 1} Nu_i = \lim_{K \to 1} Nu_a = 4 \qquad (1.139)$$

For a number of other boundary conditions, one can find friction factors and Nusselt numbers for laminar flow in annuli in the book by Shah and London [S8]. For turbulent flow in annuli and tubes, friction factors and Nusselt numbers are found in [V1] and [H3]. The formula recommended by Gnielinski in [H3] and [V1] for turbulent tube flow ($Re \geq 2300$) is

$$Nu = f \; \frac{\xi/8(Re - 1000)Pr}{1 + 12.7\sqrt{\xi/8}(Pr^{2/3} - 1)} \left[1 + \left(\frac{d_h}{L}\right)^{2/3}\right] \qquad (1.140)$$

with $\xi = \xi(Re)$ from eq. (1.124) and $f = 1$.

For annuli, the values calculated from this relation with the hydraulic diameter as a characteristic length are to be multiplied by the factors

$$f_i = 0.86 \; K^{-0.16} \text{ (inner wall, adiabatic outer wall)}$$

$$f_a = 1 - 0.14 \; K^{0.6} \text{ (outer wall, adiabatic inner wall)}$$

4 CONCLUSIONS

4.1 A Method for the Systematic Analysis of Heat Exchangers

The examples in sections 1 to 3 have demonstrated that one can systematically analyze quite different types of equipment (stirred tank, double-pipe heat exchanger, bayonet heat exchanger) applying the same method. The procedure may be set out in tabular form as a comprehensive general scheme for the solution of problems in this field:

- 1 Description of the Problem   sketch, streams, symbols
- 2 Formulation of Questions   unknowns, variables, parameters
- 3 Application of Physical Laws   balances, equilibria, kinetics
- 4 Development in Terms of the
    Unknown   eliminate, normalize, nondimensionalize
- 5 Mathematical Solution   integrate, differentiate, algebra
- 6 Discussion of the Results   limiting cases, experience, physics

---

The keywords given with each of the six items are meant as memory aids to facilitate the practical application of the method. Beginners especially are advised to follow this scheme closely when trying to solve a problem. With increasing confidence and experience, one will need it less and less.

4.2 Form of Presentation of the Results

The results of analyses of heat exchangers may be presented in a variety of forms. The presentation becomes especially convenient and comprehensive by using the definitions

$\boxed{\epsilon_i}$ — normalized temperature change of stream i (also called efficiency or effectiveness)

$\boxed{N_i}$ — number of transfer units (NTU) of stream i

$\boxed{\Theta = \epsilon_i/N_i}$ — normalized mean temperature difference,

and the functions

$$\boxed{\Theta(\epsilon_1, \epsilon_2, \text{flow configuration})} \tag{1.141}$$

$$\boxed{\Theta(N_1, N_2, \text{flow configuration})} \tag{1.142}$$

The functional relationships determined from the analysis can be given as simple formulas for the calculation of the mean temperature difference in many cases, as, e.g., $\Delta T_M = \Delta T_{LM}$ for parallel and counterflow heat exchangers. The performance of a heat exchanger depends not only on its surface area A and the overall heat transfer coefficient k that can be achieved but, to a large extent, also on the flow configuration of both fluids along the heat transfer surface. Certain idealized models of this flow configuration are characterized by the terms "stirred tank," "parallel flow," "counterflow," etc. The following chapter is devoted to the determination of the influence of flow configuration on the performance or on the efficiency ϵ or the mean temperature difference (the mean driving force) of various commonly used types of heat exchangers.

TWO

INFLUENCE OF FLOW CONFIGURATION ON HEAT EXCHANGER PERFORMANCE

1 STIRRED TANK, PARALLEL FLOW, COUNTERFLOW

1.1 Stirred Tank

The stirred tank has been treated in chapter 1 with a jacket for steam heating. Because of the unlimited specific heat capacity of condensing steam, this is a special case, with vanishing change of temperature of the vapor stream:

$$\epsilon_{\text{steam}} = C \, \epsilon_{\text{liquid in the tank}} = 0 \quad (C = 0)$$

In general, the fluid flowing through the jacket may have a finite heat capacity. Two limiting cases may be regarded.

a. The fluid in the jacket of the tank is totally mixed (as is the contents of the tank). This configuration is called "stirred tank, both sides" for brevity in the following.
b. The fluid in the jacket flows through in plug flow without any backmixing. This is called "stirred tank, one side" for short. Figure 2.1 shows these two cases schematically: stirred tank (a) both sides and (b) one side.

31

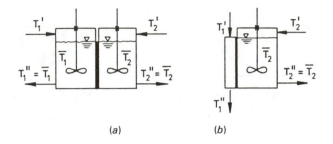

Figure 2.1 Stirred tank—heat exchanger (sketch). a(left) both sides,
b(right) one side mixed.

1.1.1 Stirred tank, both sides. In this case, the mean temperature difference between the two streams is equal to the difference of the outlet temperatures $T_1'' = \bar{T}_1$,
$T_2'' = \bar{T}_2$). We get

$$\triangle T_M = T_1'' - T_2'' \tag{2.10}$$

or

$$\Theta = \frac{T_1'' - T_2''}{T_1' - T_2'} \tag{2.2}$$

with the definitions

$$\varepsilon_1 \equiv \frac{T_1' - T_1''}{T_1' - T_2'} \qquad \varepsilon_2 \equiv \frac{T_2'' - T_2'}{T_1' - T_2'} \tag{2.3}$$

eq. (2.2) immediately leads to the relationship $\Theta(\epsilon_1, \epsilon_2)$:

$$\Theta = 1 - \varepsilon_1 - \varepsilon_2 \tag{2.4}$$

With $\epsilon_i = N_i\Theta$ (eq. [1.76], we can also find the second basic relation $\Theta(N_1, N_2)$:

$$\Theta = \frac{1}{1 + N_1 + N_2} \tag{2.5}$$

for the stirred tank, both sides. On a plot of ϵ_1 vs. ϵ_2 with Θ as a parameter [a contour
map of the function $\Theta(\epsilon_1, \epsilon_2)$], one can recognize that outlet temperatures above the
diagonal $\epsilon_1 = 1 - \epsilon_2$ —i.e., $\Theta < 0$—are impossible (see Fig. 2.2). The rays through
the origin are "balance lines" $|C| = \epsilon_2/\epsilon_1 = N_2/N_1$ (broken lines in Fig. 2.2). From
eqs. (1.76) and (2.4), it also follows that

$$\varepsilon_1 = (1 - \varepsilon_2)\frac{N_1}{1 + N_1} \qquad \text{or} \qquad \varepsilon_2 = (1 - \varepsilon_1)\frac{N_2}{1 + N_2} \tag{2.6}$$

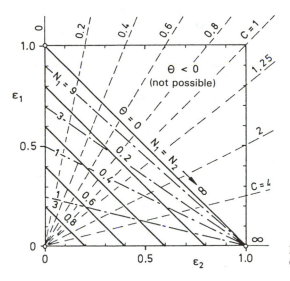

Figure 2.2 Contour map of the function $\Theta(\epsilon_1, \epsilon_2)$ for stirred tank, both sides.

The lines of constant NTU, N_1 = const. and N_2 = const. are, therefore, radii through the points (0, 1) and (1, 0), respectively, of the diagram (dash-dotted lines in Fig. 2.2). They have an intercept $N_1/(1 + N_1)$ and a slope $-N_1/(1 + N_1)$.

From diagrams like this, all relationships required for design and operation can be read off directly. The plot of Θ vs. ϵ_1, with $C = \epsilon_2/\epsilon_1$ as a parameter, fulfills the same purpose (see Fig. 2.3a). Diagrams of this kind are to be found in vol. 1 of HEDH [H3]. In this figure, the radii through the origin are lines of N_1 = const with slope $1/N_1$.

Figure 2.3b presents ϵ_1 vs. N_1, with C as a parameter. Such figures are often found in textbooks. They show distinctly the decreasing influence of transfer surface A [$N_1 = kA/\dot{M}c_p)_1$] on efficiency or on the approach to thermal equilibrium. Here the

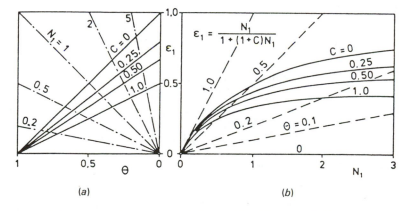

(a)

(b)

Figure 2.3 Stirred tank, both sides: a(left) $\Theta(\epsilon, C)$ and b(right) $\epsilon(N, C)$.

radii through the origin are lines $\Theta = $ const with slope Θ. Though the same information is given in the $\Theta - \epsilon$ plot of Fig. 2.3a, it covers a wider range of $N_1 \rightarrow \infty$.

1.1.2 Stirred tank, one side. The case of the stirred tank, one side, as shown schematically in Fig. 2.1b is an example of an asymmetric heat exchanger. The subscripts
1 and 2 of the two streams are not to be interchanged. The backmixed stream through
the tank is denoted in the figure by subscript 2 and the unmixed stream (plug flow) by
subscript 1. In practice, this flow configuration can be realized by a tube welded to
the outer wall of the stirred tank as a helical coil or by a coil immersed in the tank.
For the fluid in the coil, this is similar to the case of parallel or counterflow with
constant temperature on the other side $\bar{T}_2 = T_2''$. So, the mean temperature difference
is equal to the logarithmic mean

$$\triangle T_{\mathrm{M}} = \frac{(T_1' - T_2'') - (T_1'' - T_2'')}{\ln[(T_1' - T_2'')/(T_1'' - T_2'')]} \tag{2.7}$$

Contrary to the parallel flow heat exchanger, the local temperature difference at the
entrance (of stream 1) is not $T_1' - T_2'$, but $T_1' - \bar{T}_2 = T_1' - T_2''$! If $\triangle T_{\mathrm{M}}$ is divided by
the maximum difference (not actually occurring at the transfer surface) $T_1' - T_2'$ and
the changes of temperature are expressed in terms of ϵ_1 and ϵ_2 from eq. (2.3), one
obtains $\Theta(\epsilon_1, \epsilon_2)$ for this case:

$$\Theta = \frac{\varepsilon_1}{\ln[1 + \varepsilon_1/(1 - \varepsilon_1 - \varepsilon_2)]} \tag{2.8}$$

Problem: Check this formula for correctness by investigating the limiting cases
$\epsilon_2 \rightarrow 0$ (parallel and counterflow for $C \rightarrow 0$) and $\epsilon_1 \rightarrow 0$ (steamheated stirred tank
$C \rightarrow \infty$).

If the forms of presentation $\Theta(\epsilon_1, C)$ and $\epsilon_1(N_1, C)$ as in Fig. 2.3a and b are
chosen again and if we fix the notations

$$\epsilon = \epsilon_1, \ C\epsilon = \epsilon_2, \ N = N_1, \ CN = N_2$$

then the subscripts 1 or 2 can be omitted. From eq. (2.8) follows

$$\Theta = \frac{\varepsilon}{\ln\{1 + \varepsilon/[1 - (1 + C)\varepsilon]\}} \tag{2.9}$$

and, with eq. (1.76), $\epsilon = N\Theta$, one finds

$$\Theta = \left(CN + \frac{N}{1 - e^{-N}}\right)^{-1} \tag{2.10}$$

1.2 Parallel and Counterflow

The formulae for $\Theta(\epsilon_1, \epsilon_2)$ and $\Theta(N_1, N_2)$ have already been derived for parallel and counterflow heat exchangers in the first chapter. In the form $\Theta(\epsilon, C)$ and $\Theta(N, C)$ for parallel flow they are $(C \geq 0)$:

$$\Theta(\varepsilon, C) = \frac{-(1+C)\varepsilon}{\ln[1 - (1+C)\varepsilon]} \tag{2.11}$$

$$\Theta(N, C) = \frac{1 - e^{-(1+C)N}}{(1+C)N} \tag{2.12}$$

and for the counter flow $(C \leq 0)$:

$$\Theta(\varepsilon, C) = \frac{(1+C)\varepsilon}{\ln[1 + (1+C)\varepsilon/(1-\varepsilon)]} \quad (C \neq -1) \tag{2.13}$$

$$\Theta(\varepsilon) = 1 - \varepsilon \quad (C = -1)$$

$$\Theta(N, C) = \frac{1 - e^{-(1+C)N}}{N + CN\,e^{-(1+C)N}} \quad (C \neq -1) \tag{2.14}$$

$$\Theta(N) = \frac{1}{1+N} \quad (C = -1)$$

The normalized mean temperature differences Θ calculated from these equations (2.11) to (2.14) are plotted against ϵ for $-1 \leq C \leq 1$ in Fig. 2.4. Comparison with Fig. 2.3 shows that the maximum efficiency ϵ for stirred tank and parallel flow is always limited by the capacity flow rate ratio C to the value

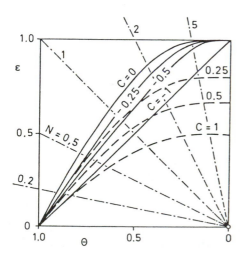

Figure 2.4 Mean temperature difference Θ as a function of change in temperature ϵ with the capacity flow rate ratio C as a parameter for parallel $(C \geq 0)$—and counterflow $(C \leq 0)$.

$$\epsilon_{max, \; st, \; p} = \frac{1}{1 + C} \tag{2.15}$$

while the maximum efficiency for counterflow is independent of C and its value is unity.

Problems

2.1 Sketch $\Theta(\epsilon)$ with $|C| = 1$ in one diagram for all flow configurations treated so far.

2.2 The following temperatures have been measured at a heat exchanger in steady operation: $T_1' = 80\,°C$, $T_2' = 20\,°C$, $T_1'' = 40\,°C$, and $T_2'' = 35\,°C$. Calculate from these ϵ ($= \epsilon_1$) and C. The flow configuration is unknown. Within what limits could $N = kA/(\dot{M}c_p)_1$ vary?

Solution: $\epsilon = 2/3$, $|C| = 3/8$, $8 \geq N \geq 1.297$.

2 CROSS FLOW

2.1 Cross Flow over One Row of Tubes (Cross Flow, One Side Mixed)

Tubular heat exchangers may be built from a single row of tubes in a rectangular frame, as shown in Fig. 2.5. Such arrangements are typically used as air coolers with the liquid in the tubes (stream 1) and air in cross flow over the tubes (stream 2). That

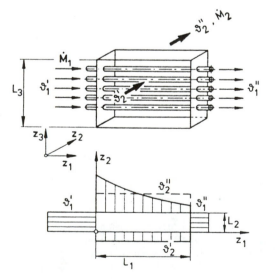

Figure 2.5 Cross flow over one row of tubes (cross flow, one side laterally mixed).

the tubes are normally finned on the outside is, however, unimportant for analysis of the influence of flow configuration on performance. The first four steps of the method of chapter 1, section 4.1, applied to the case shown in Fig. 2.5 (left to the reader as exercise), leads to a system of coupled differential equations:

$$-\frac{d\vartheta_1}{d\zeta_1} = \vartheta_1 - \bar{\vartheta}_2 \tag{2.16}$$

$$\frac{\partial\vartheta_2}{\partial\zeta_2} = \vartheta_1 - \vartheta_2 \tag{2.17}$$

Here the dimensionless length coordinates ζ_1 and ζ_2 are defined as follows (see also eq. [1.59] and [1.60] for parallel and counterflow for comparison)

$$\zeta_i \equiv \frac{kA}{(\dot{M}c_p)_i} \frac{z_i}{L_i} = N_i \cdot Z_i \qquad (i = 1, 2) \tag{2.18}$$

and $\bar{\vartheta}_2$ means

$$\bar{\vartheta}_2 = \frac{1}{N_2} \int_0^{N_2} \vartheta_2(\zeta_2)\, d\zeta_2 \tag{2.19}$$

Since the temperature of medium 1 depends only on coordinate ζ_1, while that of medium 2 depends on both, eq. (2.16) contains a total and eq. (2.17) a partial derivative. The physical significance of the equations is that the change of enthalpy in the flow direction is proportional to the heat flux and, therefore, to local temperature difference in steady state. The proportionality factors are absorbed into the expressions for the dimensionless coordinates (eq. [2.18]). For the mathematical solution (step 5 of the general method), we start by using (2.16) and (2.17) to get:

$$-\frac{d\vartheta_1}{d\zeta_1} = \frac{\partial\vartheta_2}{\partial\zeta_2} + (\vartheta_2 - \bar{\vartheta}_2) \tag{2.20}$$

This is integrated over ζ_2 from zero to N_2 for a fixed but arbitrary value of ζ_1:

$$-\int_0^{N_2} \frac{d\vartheta_1}{d\zeta_1}\, \partial\zeta_2 = \int_{\vartheta_2'}^{\vartheta_2(\zeta_1, N_2)} \partial\vartheta_2 + 0 \tag{2.21}$$

With the inlet temperatures normalized as $\vartheta_1' = 1$ and $\vartheta_2' = 0$, we get:

$$-\frac{d\vartheta_1}{d\zeta_1} N_2 = \vartheta_2(\zeta_1, N_2) \tag{2.22}$$

Similarly, eq. (2.17) can be integrated by separation of variables for a fixed value of ζ_1, i.e., also a fixed ϑ_1, with $\partial\vartheta_2 = -\partial(\vartheta_1 - \vartheta_2)$:

$$\int_{\vartheta_1 - \vartheta_2'}^{\vartheta_1 - \vartheta_2(\zeta_1, N_2)} \frac{\partial(\vartheta_1 - \vartheta_2)}{\vartheta_1 - \vartheta_2} = -\int_0^{N_2} \partial\zeta_2 \tag{2.23}$$

$$1 - \frac{\vartheta_2(\zeta_1, N_2)}{\vartheta_1(\zeta_1)} = e^{-N_2} \tag{2.24}$$

$$\vartheta_2(\zeta_1, N_2) = \vartheta_1(\zeta_1)(1 - e^{-N_2}) \tag{2.25}$$

This expression for the temperature profile of medium 2 at the outlet can be inserted now into eq. (2.22) and integrated over ζ_1:

$$-\int_1^{\vartheta_1''} \frac{d\vartheta_1}{\vartheta_1} = \frac{1 - e^{-N_2}}{N_2} \int_0^{N_1} d\zeta_1 \tag{2.26}$$

With $\vartheta_1'' = 1 - \epsilon_1$, we first obtain $\epsilon_1(N_1, N_2)$ in the form

$$\varepsilon_1 = 1 - \exp\left(-\frac{1 - e^{-N_2}}{N_2} N_1\right) \tag{2.27}$$

and so $\Theta = \epsilon_1/N_1$ as a function of N_1 and N_2 or of N and CN:

$$\Theta = \frac{1 - \exp\left(-\frac{1 - e^{-CN}}{C}\right)}{N} \tag{2.28}$$

The normalized mean temperature difference Θ can also be expressed in terms of the changes in temperature by solving for $N_2 = CN$ and using $N = \epsilon/\Theta$:

$$\Theta = \frac{-C\varepsilon}{\ln(1 + C\ln(1 - \varepsilon))} \tag{2.29}$$

For equal flow rate capacities ($C = 1$) on both sides, the maximum efficiency (for $N \to \infty$) is limited to

$$\varepsilon_{max}(C = 1) = 1 - e^{-1} = 0.632 \tag{2.30}$$

a value situated between 0.5 and 1.0, the corresponding maxima for stirred tank or parallel flow ($C = 1$) and for counterflow ($C = -1$).

2.2 Cross Flow over Several Rows of Tubes

Image n equal frames, each with one row of tubes, in the direction of flow of stream 2 in Fig. 2.5. In the tubes of each frame, n equal streams $\dot{M}_1/n$ flow in z_1 direction with a common inlet temperature $\vartheta_1' = 1$. Then the differential eqs. (2.16) and (2.17) are also valid for any one of these frames. Only, the boundaries of integration in z_2-direction are no longer 0 and N_2, but $(j - 1) N_2/n$ and jN_2/n, if we consider the jth frame ($j = 1 \ldots n$). In place of eq. (2.22), we therefore get

$$-\frac{d\vartheta_{1,j}}{d\zeta_1}\frac{N_2}{n} = \vartheta_2\left(\zeta_1,\ j\frac{N_2}{n}\right) - \vartheta_2\left(\zeta_1,\ (j-1)\frac{N_2}{n}\right) \tag{2.31}$$

and correspondingly in place of eq. (2.24):

$$\frac{\vartheta_{1,j} - \vartheta_{2,j}}{\vartheta_{1,j} - \vartheta_{2,j-1}} = e^{-N_2/n} \tag{2.32}$$

or

$$\frac{\vartheta_{2,j} - \vartheta_{2,j-1}}{\vartheta_{1,j} - \vartheta_{2,j-1}} = 1 - e^{-N_2/n} \tag{2.33}$$

Equation (2.33) can again be inserted into eq. (2.31);

$$-\frac{d\vartheta_{1,j}}{d\zeta_1} = \frac{1 - e^{-N_2/n}}{N_2/n}(\vartheta_{1,j} - \vartheta_{2,j-1}) \tag{2.34}$$

This inhomogeneous ordinary differential equation of first order has this solution:

$$\vartheta_{1,j} = e^{-g\zeta_1}\left(\vartheta_{1,j}(0) + \int_0^{\zeta_1} g\vartheta_{2,j-1}(x)\,e^{gx}\,dx\right) \tag{2.35}$$

Here the constant factor g is an abbreviation for

$$g = \frac{1 - e^{-N_2/n}}{N_2/n} \tag{2.36}$$

and the outlet temperature profile of stream 2 from the $(j-1)^{th}$ element has been denoted in short as $\vartheta_{2,j-1}$. For the first element $(j = 1)$ $\vartheta_{2,0} = \vartheta_2' = 0$, and eq. (2.35) with $\vartheta_{1,1}(0) = \vartheta_1' = 1$ yields

$$\vartheta_{1,1}(\zeta_1) = e^{-g\zeta_1} \tag{2.37}$$

For $\zeta_1 = N_1$, one gets

$$\vartheta_1'' = \vartheta_{1,1}(N_1) = e^{-gN_1} \tag{2.38}$$

which coincides with eq. (2.27) for $n = 1$. With this and eq. (2.33), the outlet temperature can be written as

$$\vartheta_{2,1}(\zeta_1) = \vartheta_{2,0} + (\vartheta_{1,1} - \vartheta_{2,0})\frac{gN_2}{n} \tag{2.39}$$

or in general

$$\vartheta_{2,j-1}(x) = \vartheta_{2,j-2}\left(1 - \frac{gN_2}{n}\right) + \vartheta_{1,j-1}\frac{gN_2}{n} \tag{2.40}$$

For $j = 2$, one obtains first from eq. (2.35),

$$\vartheta_{1,2} = e^{-g\zeta_1}\left(1 + \frac{gN_2}{n}g\zeta_1\right) \tag{2.41}$$

i.e., for $n = 2$; the mean outlet temperature of stream 1 (mixture of $\vartheta''_{1,1}$ and $\vartheta''_{1,2}$) is

$$\vartheta''_1 = 1 - \varepsilon = e^{-gN_1}\left(1 + gN_1\frac{gN_2}{4}\right)$$

$$g = \frac{1 - e^{-N_2/2}}{N_2/2} \tag{2.42}$$

For $C = 1$—i.e., $N_1 = N_2 = N$—this becomes

$$\varepsilon = 1 - e^{-gN}\left(1 + \left(\frac{gN}{2}\right)^2\right) \qquad (C = 1) \tag{2.43}$$

and the maximum efficiency is

$$\boxed{\varepsilon_{max}(n = 2, C = 1) = 1 - 2e^{-2} = 0.729} \tag{2.44}$$

i.e., a value more than 15% higher than for one row of tubes. For $j = 3$ from eq. (2.40) with $gN_2/n = b$ follows

$$\vartheta_{2,2}(x) = (1 - b)b\,e^{-gx} + (1 + b\,gx)b\,e^{-gx}$$
$$g\,e^{gx}\vartheta_{2,2}(x) = gb(2 - b + b\,gx)$$

and so

$$\vartheta_{1,3} = e^{-g\zeta_1}\left(1 + gb\left[(2 - b)\xi_1 + \frac{gb\zeta_1^2}{2}\right]\right) \tag{2.45}$$

Now, with $n = 3$ and $\vartheta''_1 = 1/3\,(\vartheta''_{1,1} + \vartheta''_{1,2} + \vartheta''_{1,3})$, we find

$$\vartheta''_1 = e^{-gN_1}\left(1 + gbN_1\left(1 - \frac{b}{3}\right) + \frac{(gbN_1)^2}{6}\right) \tag{2.46}$$

and, with $N_1 = N_2 = N$, $(gN)_\infty = n = 3$ and $b_\infty = 1$, the maximum efficiency becomes

$$\boxed{\varepsilon_{max}(n = 3, C = 1) = 1 - \frac{9}{2}e^{-3} = 0.776} \tag{2.47}$$

over 6% higher than the corresponding value for $n = 2$ rows from eq. (2.44). A similar calculation for $n = 4$ eventually leads to (problem!)

$$\varepsilon_{max}(n = 4, C = 1) = 1 - \frac{32}{3}e^{-4} = 0.805 \qquad (2.48)$$

In Appendix A, a general solution is given for an arbitrary number of tube rows n.

2.3 Ideal Cross Flow

With an increasing number of rows, larger changes of temperature, i.e., higher trans-fer performances, are achieved with the same total surface area. In the limit $n \to \infty$, i.e., for the surface area continuously distributed in z_2-direction (see cross flow plate heat exchanger in Fig. 2.6), we get the system of differential equations

$$-\frac{\partial \vartheta_1}{\partial \zeta_1} = \vartheta_1 - \vartheta_2 \qquad (2.49)$$

$$\frac{\partial \vartheta_2}{\partial \zeta_2} = \vartheta_1 - \vartheta_2 \qquad (2.50)$$

which differs by the partial derivative of ϑ_1 and the local value of ϑ_2 in eq. (2.49) from the system for cross flow, one side laterally mixed. The solution can be given in terms of power series according to Nusselt [N2]:

$$\vartheta_1(\zeta_1, \zeta_2) = \left(\sum_{n=0}^{\infty} \frac{\zeta_2^n}{n!} \sum_{m=0}^{n} \frac{\zeta_1^m}{m!} \right) e^{-(\zeta_1 + \zeta_2)} \qquad (2.51)$$

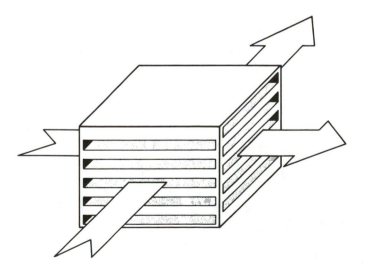

Figure 2.6 Cross flow plate heat exchanger.

$$1 - \vartheta_2(\zeta_1, \zeta_2) = \left(\sum_{n=0}^{\infty} \frac{\zeta_1^n}{n!} \sum_{m=0}^{n} \frac{\zeta_2^m}{m!} \right) e^{-(\zeta_1 + \zeta_2)} \tag{2.52}$$

Writing out the first three terms of eq. (2.51), one can recognize the structure of these solutions and check their validity with some limiting cases:

$$\vartheta_1 = \left(1 + (1 + \zeta_1)\zeta_2 + \left(1 + \zeta_1 + \frac{\zeta_1^2}{2!} \right) \frac{\zeta_2^2}{2!} + \cdots \right) e^{-(\zeta_1 + \zeta_2)}$$

The corresponding solution for $1 - \vartheta_2$ follows by interchanging the subscripts 1 and 2. For $\zeta_1 = 0$ and arbitrary values of ζ_2, $\vartheta_1 = \vartheta_1' = 1$ must be valid. One finds, in this case,

$$\vartheta_1(0, \zeta_2) = \sum_{n=0}^{\infty} \frac{\zeta_2^n}{n!} e^{-\zeta_2} \tag{2.53}$$

which, in fact, gives the value 1 as $e^x = \Sigma x^n / n!$ At the "inlet edges" of the heat exchanger, the respective temperatures of the other side are constant. Therefore, for these flow paths, the equations (2.49) and (2.50) can be integrated directly by separation of variables. We get

$$\vartheta_1(\zeta_1, 0) = e^{-\zeta_1}$$

and

$$1 - \vartheta_2(0, \zeta_2) = e^{-\zeta_2} \tag{2.54}$$

which is also fulfilled by eqs.(2.51) and (2.52), as can be easily seen from the "written out" form of eq. (2.51). Analogous to the mixture of partial streams from parallel rows of tubes, the mean outlet temperatures are now obtained by integration of the local variations of temperature along the "outlet edges" of the heat exchanger:

$$\vartheta_1'' = \frac{1}{N_2} \int_0^{N_2} \vartheta_1(N_1, \zeta_2) \, d\zeta_2$$

$$\vartheta_2'' = \frac{1}{N_1} \int_0^{N_1} \vartheta_2(\zeta_1, N_2) \, d\zeta_1 \tag{2.55}$$

The result of this integration can be written with $1 - \vartheta_1'' = \epsilon$, $\vartheta_2'' = C\epsilon$, $N_2 = CN$ and $\epsilon = N\Theta$, in the form $\Theta(N, C)$:

$$\Theta = \sum_{m=0}^{\infty} \frac{1 - \sum_{n=0}^{m} e^{-N} N^n / n!}{N} \frac{1 - \sum_{n=0}^{m} e^{-CN} (CN)^n / n!}{CN} \tag{2.56}$$

Figure 2.7 shows a short computer program to evaluate this formula. It may be surprising that only one loop of the program is necessary, in spite of the "box-in-a-box" double summation (inner sums from 0 to m, outer summation over m). The summation ends when the mth term p adds an amount less than δP to the sum P. For $\delta = 10^{-5}$, $m = 5$ terms are sufficient for $(N, C) = (1, 1)$, and one finds $\epsilon(1, 1) = 0.4762$; $m = 119$ terms are required, however, for $(N, C) = (100, 1)$ to obtain the same accuracy [$\epsilon(100, 1) = 0.9436$]. For equal flow capacities, $C = 1$, and large NTU, the efficiency may be obtained easier from the asymptotic formula [V1]

$$ \epsilon = 1 - \frac{1}{\sqrt{\pi N}}\left(1 - \frac{1}{16N}\right) \qquad (N > 2) \qquad (2.57) $$

the unmixed or ideal cross flow gives, with eq. (2.57) for $C = 1$, a maximum efficiency of $\epsilon_{max} = 1$ that has been found only for counterflow from the analyses done so far. An efficiency of 90% ($\epsilon = 0.9$) is reached for ideal cross flow only with $N = 32$ (at $C = 1$). For counterflow, however, $N = 9$ [$\epsilon(C = -1) = N/(1 + N)$] is sufficient. The transfer surface of the crossflow exchanger would have to be more than 3.5 times larger than that of an ideal counterflow exchanger to reach an efficiency of 90% under the same conditions.

Having reached this state of perception, a beginner is inclined to ask why manufacturers of heat exchangers can offer, and are obviously able to sell, apparatuses with flow configurations other than counterflow. It may be worthwhile to think about this and to investigate the perceptions and assumptions underlying this question in detail.

2.4 Cross Flow, Both Sides Laterally Mixed

In section 2.1, we first treated the asymmetric case of crossflow one side laterally mixed; then, in the next section, the progressively more favorable and less asymmetric case of crossflow over increasing numbers of rows of tubes; and, finally, the symmetric ideal crossflow. One could also imagine another symmetric case: crossflow with both sides laterally mixed. Mathematically, this case differs from the preceding one by total in place of the partial derivatives in the basic equations:

$$ -\frac{d\vartheta_1}{d\zeta_1} = \vartheta_1 - \bar{\vartheta}_2 \qquad (2.58) $$

$$ \frac{d\vartheta_2}{d\zeta_2} = \bar{\vartheta}_1 - \vartheta_2 \qquad (2.59) $$

ϑ_1 and, thus, $d\vartheta_1/d\zeta_1$ depend only on ζ_1, while ϑ_2 and $d\vartheta_2/d\zeta_2$ only on ζ_2. Thus, on the right hand side of these two equations, we have to use the integral mean values of the other temperature:

$$ \bar{\vartheta}_1 = \frac{1}{N_1}\int_0^{N_1} \vartheta_1(\zeta_1)\,d\zeta_1 \qquad (2.60) $$

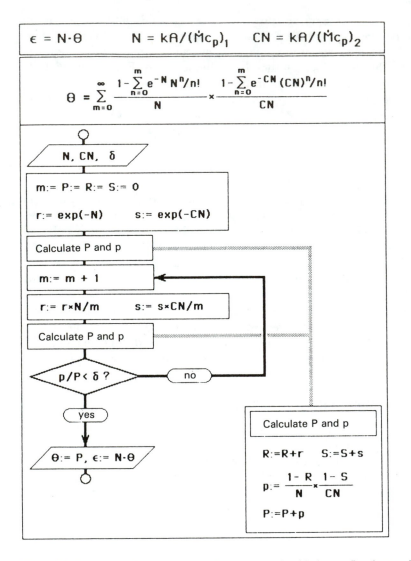

Figure 2.7 Program for the calculation of the efficiency of an ideal cross flow heat exchanger.

$$\bar{\vartheta}_2 = \frac{1}{N_2} \int_0^{N_2} \vartheta_2(\zeta_2)\, d\zeta_2 \tag{2.61}$$

Both eqs. (2.58) and (2.59) can be easily integrated:

$$\frac{\vartheta_1 - \bar{\vartheta}_2}{1 - \bar{\vartheta}_2} = e^{-\zeta_1} \tag{2.62}$$

$$\frac{\bar{\vartheta}_1 - \vartheta_2}{\bar{\vartheta}_1 - 0} = e^{-\zeta_2} \tag{2.63}$$

The mean values defined above follow from this to be

$$\frac{\bar{\vartheta}_1 - \bar{\vartheta}_2}{1 - \bar{\vartheta}_2} = \frac{1 - e^{-N_1}}{N_1} \tag{2.64}$$

$$\frac{\bar{\vartheta}_1 - \bar{\vartheta}_2}{\bar{\vartheta}_1} = \frac{1 - e^{-N_2}}{N_2} \tag{2.65}$$

These two equations can be simultaneously solved for $\bar{\vartheta}_1$ and $\bar{\vartheta}_2$ (the details of which are left as an exercise to the reader). From eqs. (2.62) and (2.63) with $\zeta_1 = N_1$, $\vartheta_1'' = \vartheta_1(N_1)$, $1 - \vartheta_1'' = \epsilon$ and $\Theta = \epsilon/N$, we finally obtain

$$\Theta = \left(\frac{N}{1 - e^{-N}} + \frac{CN}{1 - e^{-CN}} - 1 \right)^{-1} \tag{2.66}$$

This result is remarkable insofar as the maximum efficiency calculated from it is not obtained for $N \to \infty$ as in all other cases treated so far, but for a finite value of $N \approx 3$:

$$\varepsilon_{\max}(C = 1, N = 3) = 0.5645 \tag{2.67}$$

The transfer performance is a maximum at a certain transfer surface area corresponding to $N = 3$. For larger surface areas than this value, the performance will decrease again to reach the same limiting value as found earlier for stirred tank and parallel flow:

$$\varepsilon_\infty = \frac{1}{1 + C} \tag{2.68}$$

2.5 Comparison of Simple Flow Configurations

A comparison of the simple flow configurations treated so far can be shown most clearly on a plot of ϵ vs. Θ for equal flow capacity rates on both sides, $|C| = 1$ (Fig. 2.8). Here again, the rays through the origin are lines of constant NTU. The range of crossflow configurations is depicted in gray. It is intermediate between parallel flow and counterflow. The diagonal $N = 1$ subdivides the figure into two typical ranges. At low NTU, $N < 1$, the flow configuration, apart from backmixing, has little influence on transfer performance. Here the heat transfer coefficients and the surface area are the important parameters. At high NTU, $N > 1$, flow configuration becomes crucial, while increasing surface area has little or even a negative influence in some cases. To reach efficiencies of, say, 80% or more, only ideal crossflow and counterflow come into reckoning. In the limiting case $C = 0$ (constant temperature on the other side, see broken curve in Fig. 2.8), flow configuration plays no role. This curve

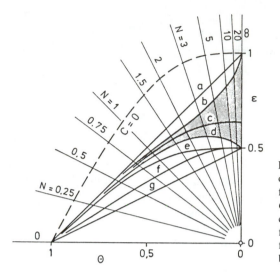

Figure 2.8 Influence of flow configuration on transfer performance at equal capacity flow rates $|C| = 1$ ————— and for $C = 0$ —————— **a** counterflow; **b** ideal cross flow; **c** cross flow, one side; **d** cross flow both sides laterally mixed; **e** parallel flow; **f** stirred tank, one side; **g** stirred tank, both sides.

is valid for all configurations with the exception of "stirred tank, both sides." The curve for $C = 0$ for the stirred tank coincides with the diagonal, i.e., with the one for counterflow at $C = -1$. Not shown in the figure for clarity are the lines for finite numbers of tube rows which lie between crossflow, one side mixed, and ideal crossflow.

Problems

2.3 Obtain from eq. (2.56) the normalized mean temperature difference Θ for ideal crossflow as a function of N in the limiting case $C \to 0$. Compare the result with those for parallel flow and counterflow!

Solution: same as in Problem 1.6.

2.4 Draw a diagram of ϵ vs. N for $|C| = 1$ corresponding to Fig. 2.8 with $0 \le N \le 10$.

2.5 What is the outlet temperature $\vartheta''_{1,2}$ attained by the second partial stream ($\dot{M}_1/2$) with crossflow over two rows, if it flows in reverse ($-z_1$) direction? What value of ϵ_{max} ($C = 1$) would be obtained with that configuration?

3 COMBINED FLOW CONFIGURATIONS

3.1 Cross-Counterflow

In crossflow over tubes and tube bundles, one can usually get higher heat transfer coefficients than in longitudinal flow. This compensates, to a certain extent, for the lower efficiencies of crossflow as compared to counterflow. One possibility of increasing the efficiency beyond that of pure crossflow for the simple crossflow elements shown in Fig. 2.5 is to put such elements together in a counterflow arrangement as shown in Fig. 2.9. Three versions of cross-counterflow configuration are shown in this figure. In the counterflow cascade (Fig. 2.9a), the crossflow elements are so connected by tubing that the stream M_1 may be regarded as laterally completely mixed between the units of the cascade. Obviously, in counter-directional (Fig. 2.9b) and co-directional cross-counterflow (Fig. 2.9c), the stream M_1 is not laterally mixed between the elements. In case a, the four units or "cells" are coupled only through the mean outlet temperatures. Such a cascade is analyzed most easily. With N and C known for each cell, the inlet and outlet temperatures of each cell are related by the function $\epsilon(N, C)_{cell}$. For the cell with number j, we have

$$\frac{\vartheta'_{1j} - \vartheta''_{1j}}{\vartheta'_{1j} - \vartheta'_{2j}} = \varepsilon_j(N_j, C_j) \tag{2.69}$$

and

$$\frac{\vartheta''_{2j} - \vartheta'_{2j}}{\vartheta'_{1j} - \vartheta'_{2j}} = |C_j|\varepsilon_j(N_j, C_j) \tag{2.70}$$

It is convenient to reduce the number of subscripts and represent the temperature of the two streams thus

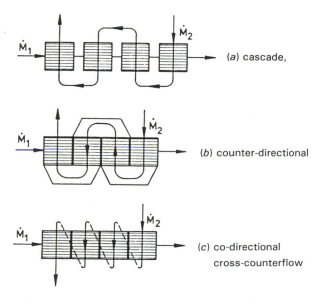

Figure 2.9 Cross-counterflow configurations: **a** counterflow cascade; **b** counter-directional; **c** co-directional cross-counterflow.

$$\Theta := \vartheta_1 \tag{2.71}$$

$$\vartheta := \vartheta_2 \tag{2.72}$$

The subscripts now used for the temperatures pertain to the cells (see Fig. 2.10). In place of $|C|$, we now write R and solve for the outlet temperatures in cell j as

$$\boxed{\Theta''_j = (1 - \varepsilon_j)\Theta'_j + \varepsilon_j\vartheta'_j} \tag{2.73}$$

and

$$\boxed{\vartheta''_j = R_j\varepsilon_j\Theta'_j + (1 - R_j\varepsilon_j)\vartheta'_j \qquad (j = 1,\dots,J)} \tag{2.74}$$

These give $2J$ equations for the $4J$ unknown temperatures of the cascade. $2J$ more equations required are found form the interconnections of the cells. If the cells in Fig. 2.9a are numbered following the direction of $\dot{M}_1$ by $j = 1$ to 4, we get

$$\Theta'_1 = 1, \quad \Theta'_2 = \Theta''_1, \quad \Theta'_3 = \Theta''_2, \quad \Theta'_4 = \Theta''_3,$$
$$\vartheta'_1 = \vartheta''_2, \quad \vartheta'_2 = \vartheta''_3, \quad \vartheta'_3 = \vartheta''_4, \quad \vartheta'_4 = 0$$

or, in general, for countercurrent cascades:

$$\Theta'_j = \Theta''_{j-1} \quad \text{with} \quad \Theta''_0 = \Theta' = 1 \quad (\text{Inlet } \dot{M}_1) \tag{2.75}$$

and

$$\vartheta'_j = \vartheta''_{j+1} \quad \text{with} \quad \vartheta''_{J+1} = \vartheta' = 0 \quad (\text{Inlet } \dot{M}_2) \tag{2.76}$$

The last two (or $2J$) equations can be inserted directly into eqs. (2.73) and (2.74) in order to eliminate all $2J$ cell inlet temperatures (single primed):

$$\Theta''_j = (1 - \varepsilon_j)\Theta''_{j-1} + \varepsilon_j\vartheta''_{j+1} \tag{2.77}$$

$$\vartheta''_j = R_j\varepsilon_j\Theta''_{j-1} + (1 - R_j\varepsilon_j)\vartheta''_{j+1} \tag{2.78}$$

For the example shown in Fig. 2.9, this is a system of 8 linear equations with 8 unknowns. The system of equations can be solved easily by introducing a ratio of

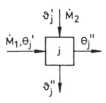

Figure 2.10 Cell of a cascade.

difference of temperatures before and after a cell for the streams M_1 and M_2, derived from eqs. (2.73) and (2.74) as

$$\frac{\delta_{j0}}{\delta_{jl}} = \frac{\Theta'_j - \vartheta''_j}{\Theta''_j - \vartheta'_j} = \frac{1 - R_j \varepsilon_j}{1 - \varepsilon_j} \tag{2.79}$$

Based on this definition of δ and with eqs. (2.75) and (2.76), we get

$$\delta_{j+1.0} = \Theta'_{j+1} - \vartheta''_{j+1} = \Theta''_j - \vartheta'_j = \delta_{j,l} \tag{2.80}$$

We can now elegantly eliminate all intermediate temperatures by multiplying these ratios for cells 1 to 4:

$$\frac{1 - \vartheta''}{\Theta''_1 - \vartheta'_1} \frac{\Theta'_2 - \vartheta''_2}{\Theta''_2 - \vartheta'_2} \frac{\Theta'_3 - \vartheta''_3}{\Theta''_3 - \vartheta'_3} \frac{\Theta'_4 - \vartheta''_4}{\Theta'' - 0} = \prod_{j=1}^{4} \frac{1 - R_j \varepsilon_j}{1 - \varepsilon_j}$$

Defining ε as $1 - \Theta''$ and $R\varepsilon$ as ϑ'', we can express the relationship between the individual cell efficiencies ε_j and the efficiency of the whole cascade ε in similar form:

$$\boxed{\frac{1 - R\varepsilon}{1 - \varepsilon} = \prod_{j=1}^{J} \frac{1 - R_j \varepsilon_j}{1 - \varepsilon_j}} \tag{2.81}$$

No statement whatsoever has been made on the flow configuration inside the cells, so that the result is valid for countercurrent cascades of arbitrary heat exchanger cells. For constant heat capacity $R_j = R$ and for J identical cells this yields

$$\frac{1 - R\varepsilon}{1 - \varepsilon} = \left(\frac{1 - R\varepsilon_j}{1 - \varepsilon_j}\right)^J \tag{2.82}$$

These equations are also correct in the limiting case $C \to -1$ or $R = |C| \to 1$; they do not yield a value of ε, however, but only the identity $1 = 1^J$. By calculating the limiting value for $(1 - R) \to 0$ one can find (exercise!)

$$\boxed{\frac{\varepsilon}{1 - \varepsilon} = \sum_{j=1}^{J} \frac{\varepsilon_j}{1 - \varepsilon_j} \qquad (R = 1)} \tag{2.83}$$

i.e., four equal cells with $\varepsilon_j = 0.5$ in a countercurrent cascade give an overall efficiency of $J/(1 + J)$, that means 80% for $J = 4$.

Cases *b* and *c* of Fig. 2.9 have to be treated in a different way, as the elements are coupled here through each row of tubes. For example, one can subdivide each element into $K \times J$ smaller cells (say 10×10), and again apply eqs. (2.73) and (2.74), together with the corresponding relations from the interconnections of these cells (e.g., 800 linear equations with 800 unknowns). The solution is found most easily by an iterative procedure because of the sparsely occupied matrices as each single equation contains only three unknowns. Figure 2.11 shows efficiencies ε vs. NTU with equal flow capacities on both sides ($R = |C| = 1$) calculated by the above procedure for co-directional and counter-directional fourfold cross-counterflow.

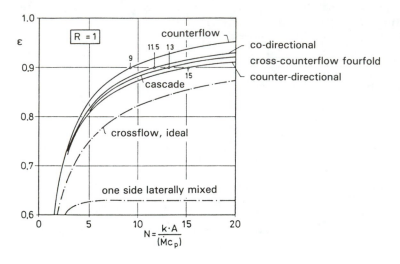

Figure 2.11 Efficiency of cross-counterflow configurations.

It may be seen from the figure that the co-directional cross-counterflow gives somewhat higher and the counter-directional somewhat lower efficiencies than the cascade of four ideal crossflow cells. The NTU-values required to reach $\epsilon = 0.9$ are marked in this figure for pure counterflow ($N = 9$), co-directional cross-counterflow ($N = 11.5$), the countercurrent cascade ($N = 13$), and the counter-directional cross-counterflow ($N = 15$). For ideal crossflow, the corresponding value ($N = 32$) lies outside the diagram. With tube bundles of rectangular cross-section (built from elements as shown schematically in Fig. 2.5), counter-directional cross-counterflow is technically easier to realize than co-directional. One only needs simple turnaround headers over each pair of single elements, as shown in Fig. 2.9b. Co-directional cross-counterflow, however, would require long turnaround ducts with correspondingly higher pressure drop and larger heat losses. In practice, one will, therefore, content oneself in these cases with the less favorable configuration ($N_{90\%} = 15$). The more favorable co-directional cross-counterflow can be realized, however, by a quite different design [S7, M4]: The tubes are wound in several concentric layers to form multiple coils in a very compact apparatus (see Fig. 2.12 *a, b*), coming very close to ideal counterflow.

Problems

2.6 Corresponding to eq. (2.81), derive a relationship for cross-cocurrent flow!

Solution: $1 - (1 + R)\epsilon = \Pi[1 - (1 + R_j)\epsilon_j]$

2.7 What is the efficiency of a cocurrent cascade with $R = R_j = 1$ and $\epsilon_j = 0.95$ for $J = 2, 3, 4, 5, 6$?

Figure 2.12 Wound multiple-coil heat exchanger as an example for co-directional cross-counterflow (photo: A. G. Linde).

Solution: (0.095, 0.865, 0.172, 0.795, 0.234)

Sketch the graph of $\epsilon(J)$! Determine the asymptotic value $\lim_{J \to \infty} \epsilon(J)$!

2.8 In the Linde air liquefaction process [H1] air from the surroundings has to be compressed and cooled, e.g., from 20 °C down to -180 °C, so that, after expansion, about 6% of the mass flow can be led off as a liquid. The remaining 94% of the expanded air, now at about -194 °C, is used to cool the incoming air in a co-directional multiple cross-counterflow heat exchanger (as in Fig. 2.12). This air is thereby heated up to 19 °C. Calculate the efficiency ϵ, the capacity ratio C, and the number of transfer units N required (based on the outlet air stream). The number of turns of the coils may be large enough to reach practically ideal counterflow.

Solution: $\epsilon = 0.995$, $C = -0.939$, $N = (213/13)\ln 14 = 43.24$

For comparison, calculate the efficiency ϵ reached with that N in a countercurrent cascade of 50 identical crossflow elements, one side laterally mixed.

Solution: $N_j = 0.8648$, $\epsilon_j = 0.447$, $\epsilon = 0.994$.

3.2 Shell-and-Tube Heat Exchangers with Baffles

Figure 2.13 shows a shell-and-tube heat exchanger having two tube-side passes and two baffles. On the shell side, this results in three compartments where the fluid is essentially in crossflow to the tubes. For the purpose of analysis, the apparatus is subdivided into six cells as shown in Fig. 2.14. For each cell we can again write down the eqs. (2.73) and (2.74) [G1–G3]. The interconnection of inlet and outlet temperatures of subsequent cells may now be written in the following form:

$$\Theta'_j = \Theta''_{j-1} \tag{2.84}$$

$$\vartheta'_j = \vartheta''_k \tag{2.85}$$

Here j is the cell number along the direction of stream $\dot{M}_1$ (used with Θ) and k is the number of the cell lying upstream of j in the direction of $\dot{M}_2$ (used with ϑ). Once the flow configuration is fixed, each k is related unequivocally to a certain j. In the example of Fig. 2.14, we find the following relation:

j	1	2	3	4	5	6
$k(j)$	6	3	4	0	2	5

The linear system of equations for the $2J$ unknown outlet temperatures becomes

$$\Theta''_j = (1 - \varepsilon_j)\Theta''_{j-1} + \varepsilon_j \vartheta''_k \tag{2.86}$$

$$\vartheta''_j = R_j \varepsilon_j \Theta''_{j-1} + (1 - R_j \varepsilon_j)\vartheta''_k \tag{2.87}$$

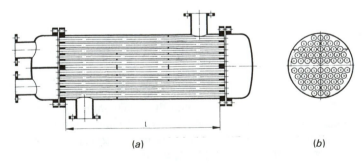

(a)   (b)

Figure 2.13 Shell-and-tube heat exchanger with two tube-side passes and two baffles **a** longitudinal, **b** cross section.

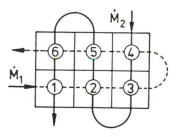

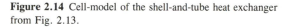

Figure 2.14 Cell-model of the shell-and-tube heat exchanger from Fig. 2.13.

It can be solved using the well-known methods for systems of linear equations. For all the flow configurations with two passes on one side (as, for example, in Figs. 2.12 and 2.13), the problem can be solved explicitly by an elegant method recently shown by Gaddis and Vogelpohl [G4]. As in Fig. 2.15, the configuration of cells is first drawn in a straight sequence with the corresponding flow paths of the two streams. One can then easily recognize that the cells 2, 3, and 4 form a countercurrent cascade, while cells 5 and 6 are connected in parallel flow.

Using the formulae for the efficiency of a parallel and a counterflow arrangement of two cells, as compiled in Fig. 2.16, the combined efficiencies of the new "double-cells" (2–3) [(2–3)–4], and (5–6) can be easily calculated from the efficiencies of the individual cells. For example, with $\epsilon_j = 0.4$ ($j = 1, \ldots ,6$) and $R = 1$, we find:

$$\varepsilon_{2.3} = \frac{2\varepsilon_j}{1 + \varepsilon_j} \qquad\qquad = 0.571$$

$$\varepsilon_{5.6} = 2\varepsilon_j(1 - \varepsilon_j) \qquad\qquad = 0.480$$

$$\varepsilon_{2.4} = \frac{\varepsilon_{2.3} + \varepsilon_j - 2\varepsilon_{2.3}\varepsilon_j}{1 - \varepsilon_{23}\varepsilon_j} \qquad\qquad = 0.667$$

$$\varepsilon_{2.6} = \varepsilon_{2.4} + \varepsilon_{5.6} - 2\varepsilon_{2.4}\varepsilon_{5.6} \qquad = 0.507$$

$$\varepsilon = \frac{\varepsilon_j + \varepsilon_{2.6} - 2\varepsilon_j\varepsilon_{2.6}}{1 - \varepsilon_j\varepsilon_{2.6}} \qquad\qquad = 0.629$$

By changing the direction of stream 1, stream 2, or both, three other flow configurations can be realized, for each of which the overall efficiency can be calculated as above (exercise!). Changing the direction of either stream 1 or stream 2 leads to a lower efficiency of $\epsilon = 0.540$ in both cases. With the directions of both streams reversed, there is no change in the efficiency: $\epsilon = 0.629$. The four possible flow configurations, thus, fall into two pairs which yield the same efficiency but are by no means identical. This may be verified from a sketch; one transposes to the other when both flow directions are reversed. Following Gaddis [G1–G4], we distinguish the four possible flow configurations by the distance between the two inlet cells and

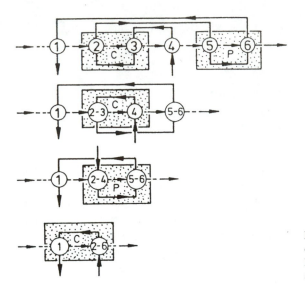

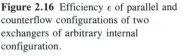

Figure 2.15 How to determine the overall efficiency of the shell-and-tube heat exchanger from Fig. 2.13, according to Gaddis a. Vogelpohl.

denote them by G1, G2, G3, and G4 (see Fig. 2.17). In the case of the present geometry (2 × 3 cells), these four possibilities can be realized with one single apparatus just by changing the flow directions. The example in Fig. 2.14 corresponds to configuration G4. The two configurations with the same efficiency exhibit different internal variations of temperature, as can be verified by calculation. Figure 2.17 shows the variation of temperature of both streams for all four configurations, calcu-

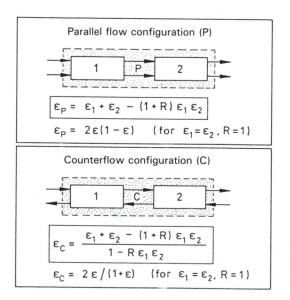

Parallel flow configuration (P)

$$\varepsilon_P = \varepsilon_1 + \varepsilon_2 - (1+R)\,\varepsilon_1\varepsilon_2$$
$$\varepsilon_P = 2\varepsilon(1-\varepsilon) \quad (\text{for } \varepsilon_1 = \varepsilon_2, R = 1)$$

Counterflow configuration (C)

$$\varepsilon_C = \frac{\varepsilon_1 + \varepsilon_2 - (1+R)\,\varepsilon_1\varepsilon_2}{1 - R\,\varepsilon_1\varepsilon_2}$$
$$\varepsilon_C = 2\varepsilon/(1+\varepsilon) \quad (\text{for } \varepsilon_1 = \varepsilon_2, R = 1)$$

Figure 2.16 Efficiency ϵ of parallel and counterflow configurations of two exchangers of arbitrary internal configuration.

lated with the data used above ($\epsilon_j = 0.4$, $R = 1$, 6 cells). The temperatures for G1 and G2 change monotonically along the flow direction, while relative maxima (or minima) occur in G3 and G4!

It is, therefore, by no means unimportant how such an apparatus is ducted for the two streams. A temperature sensitive fluid, which is heated to a desired outlet temperature might have passed inadmissibly high temperatures internally. At equal—logically the higher—efficiency, the configuration G2 would be preferable here.

The temperature scales in Fig. 2.17 are labelled only with the value 0.5. The inlet temperatures of either stream may be put equal to one or to zero arbitrarily.

Using the ϵ-values of the combined groups of cells from Fig. 2.15, one can also calculate the intermediate temperatures. If two temperatures are known at a cell or a group of cells, then the other two can be calculated via ϵ and $R\epsilon$. This can be done in the inverse sequence of the calculation of the overall efficiency (our outlet temperature). Corresponding to the last scheme in Fig. 2.15, one can first calculate the outlet temperature Θ_1'' of stream 1 from the definition of efficiency for group (2-6):

$$\varepsilon_{2.6} = \frac{\varepsilon - \Theta_1''}{1 - \Theta_1''} = \frac{1 - \vartheta_6''}{1 - \Theta_1''}$$

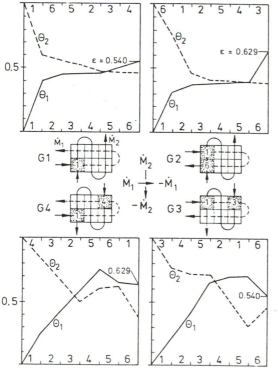

Figure 2.17 Temperature variations for the four possible configurations of the apparatus from Fig. 2.13.

$$\Theta_1'' = \frac{\varepsilon - \varepsilon_{2,6}}{1 - \varepsilon_{2,6}} \qquad\qquad = 0.247$$

$$\vartheta_6'' = 1 - (1 - \Theta_1'')\varepsilon_{2,6} \qquad\qquad = 0.618$$

For group (2–4), we find:

$$\varepsilon_{2,4} = \frac{\Theta_4'' - \Theta_1''}{1 - \Theta_1''} = \frac{1 - \vartheta_2''}{1 - \Theta_1''}$$

$$\Theta_4'' = \Theta_1'' + (1 - \Theta_1'')\varepsilon_{2,4} \qquad\qquad = 0.749$$

$$\vartheta_2'' = 1 - (1 - \Theta_1'')\varepsilon_{2,4} \qquad\qquad = 0.498$$

For cell 4, one obtains:

$$\varepsilon_4 = \frac{\Theta_4'' - \Theta_3''}{1 - \Theta_3''} = \frac{1 - \vartheta_4''}{1 - \Theta_3''}$$

$$\Theta_3'' = \frac{\Theta_4 - \varepsilon_j}{1 - \varepsilon_j} \qquad\qquad = 0.581$$

$$\vartheta_4'' = 1 - (1 - \Theta_3'')\varepsilon_j \qquad\qquad = 0.832$$

It follows for cell 3:

$$\varepsilon_3 = \frac{\Theta_3'' - \Theta_2''}{\vartheta_4'' - \Theta_2''} = \frac{\vartheta_4'' - \vartheta_3}{\vartheta_4'' - \Theta_2''}$$

$$\Theta_2'' = \frac{\Theta_3'' - \varepsilon_j \vartheta_4''}{1 - \varepsilon_j} \qquad\qquad = 0.413$$

$$\vartheta_3'' = \vartheta_4'' - (\vartheta_4'' - \Theta_2'')\varepsilon_j \qquad\qquad = 0.665$$

And, correspondingly, for cell 5:

$$\varepsilon_5 = \frac{\Theta_5'' - \Theta_4''}{\vartheta_2'' - \Theta_4''} = \frac{\vartheta_2'' - \vartheta_5''}{\vartheta_2'' - \Theta_4''}$$

$$\Theta_5'' = \Theta_4'' + (\vartheta_2'' - \Theta_4'')\varepsilon_j \qquad\qquad = 0.649$$

$$\vartheta_5'' = \vartheta_2'' - (\vartheta_2'' - \Theta_4'')\varepsilon_j \qquad\qquad = 0.598$$

Now all temperatures are known. Finally, one can check for cell 6, whether

$$\frac{\varepsilon - \Theta_5''}{\vartheta_5'' - \Theta_5''} = \varepsilon_j = \frac{\vartheta_5'' - \vartheta_6''}{\vartheta_5'' - \Theta_5''}$$

is fulfilled. We find 0.392 in place of 0.400, i.e., an error of 2%, due to the fact that we did the calculations with only three digits. The error could be reduced by storing the intermediate values with a larger number of digits. The outlet temperature Θ_6'' is equal to the total efficiency $\varepsilon = 0.629$; and, for ϑ_1'', we get $1 - \varepsilon = 0.371$. For configurations with more than two passes on one side, it is more convenient to solve the linear system of eqs. (2.86) and (2.87) numerically. As input data, one only needs the cell efficiencies ε_j and $(Re)_j$ and the relation of the numbers $k(j)$ characterizing the configuration. Most easily programmed, and probably most economical with respect to storage capacity required, is the simple iterative procedure in which all Θ_j'' and all ϑ_j'' are first set equal to zero and one, respectively (or vice versa, depending on the choice of normalization). The eqs. (2.86) and (2.87) then yield first approximations for Θ_j'' and ϑ_j''; these are again inserted on the right hand sides and so forth until the difference between two successive steps of iteration becomes smaller than a specified threshold value of accuracy.

Figure 2.18 shows a flowsheet of a short computer program that may be implemented even on a pocket calculator (it is universally valid for any one of the four configurations G1 to G4 and is considerably simpler than the one suggested by Gaddis in [G3] that requires a special subroutine for each configuration). Calculations with programs like this show that the efficiency for exchangers with a fixed cell number in one direction (e.g., two passes on one side) and increasing number of cells in the other direction tend to a common asymptotic value either in a monotonic or an alternating sequence, when the number of transfer units is kept constant (see Fig. 2.19).

The cell efficiency of $\varepsilon_j = 0.4$ used in the numerical examples so far may be obtained, for instance, with crossflow one side laterally mixed with $J = 6$; $N_j = 0.715$; and, therefore, $N = 6N_j = 4.29$. With one baffle, i.e., a total of four cells, one finds $N_j = 4.29/4$, $\varepsilon_j = 0.482$. The efficiencies of the four possible configurations are $\varepsilon = 0.509$ (for G1 and G4) and $\varepsilon = 0.659$ (for G2 and G3) (easily checked with the method from Fig. 2.15). With only two crossflow cells, one side mixed, the parallel and the countercurrent arrangement give (with $N_j = 4.29/2$, $\varepsilon_j = 0.586$) the efficiencies $\varepsilon = 0.485$ (for G1 = G3) and $\varepsilon = 0.739$ (for G2 = G4). With three baffles, i.e., eight cells, one obtains, with $N_j = 4.29/8$, $\varepsilon_j = 0.3397$, the efficiencies $\varepsilon = 0.556$ (for G1 and G4) and $\varepsilon = 0.613$ (for G2 and G3).

These results are plotted in Fig. 2.19 against $J/2$, which may be termed as the half cell number ($J/2$ = number of baffle spaces = number of baffles + 1). Similar diagrams, treating the cells as "stirred tanks," can be found in Gaddis and Schlünder [G3]. The more favorable cases—G3 with even number of half-cells, G4 with odd

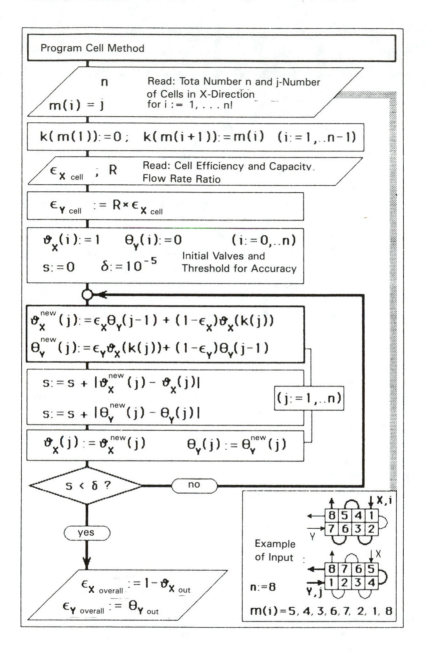

Figure 2.18 Flow sheet of a simple program for numerical calculation of temperature variations in shell-and-tube heat exchangers.

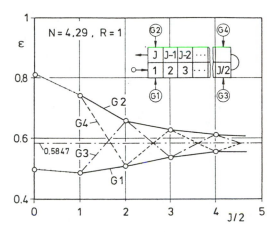

Figure 2.19 Overall efficiency of a two-pass shell-and-tube heat exchanger as a function of the number of cells and the type of configuration.

number of half-cells, or G2—have a decreasing efficiency with an increasing number of cells. For the less favorable cases—G3 with odd number, and G4 with even number of half-cells, or G1—the converse is true. For comparison, the cases of pure parallel and counterflow ($\epsilon = 0.499$ and $\epsilon = 0.811$, respectively) are shown in the figure at $J = 0$.

The difference between the efficiency of an apparatus of either type with three baffles ($J/2 = 4$) and the common asymptotic efficiency value is only about $\pm 5\%$. The common asymptote is obviously characterized by the fact that stream 2 becomes more and more an axial stream (parallel or antiparallel to stream 1), which is mixed laterally to the main direction due to the baffles. This limiting case may again be analyzed similar to the double-pipe bayonet heat exchanger, found in chapter 1, section 3 (problem!). The expression for $\epsilon(N, R)$ as already shown by Underwood [U1] is

$$\epsilon = \frac{2(1 - e^{-\omega N})}{(\omega + 1 + R) + (\omega - 1 - R)e^{-\omega N}} \qquad \omega = \sqrt{1 + R^2} \qquad (2.88)$$

With $R = 1$ and $N = 4.29$, the asymptotic value of ϵ is found from eq. (2.88) as 0.5847, shown in Fig. 2.19. For $R = 0$, the formula yields $\epsilon = 1 - e^{-N}$, a result obtained earlier and valid for all flow configurations without backmixing. The maximum value for $N \to \infty$ is

$$\epsilon_\infty = \frac{2}{\omega + 1 + R} \qquad (2.89)$$

For $R = 1$, it gives $\epsilon = 2/(2 + \sqrt{2}) = 0.5858$, a value which had already been very closely approached by $\epsilon = 0.5847$ for $N = 4.29$. Equation (2.88) can also be solved inversely for N, so that one can get the relation $\Theta(\epsilon, R)$ with $N = \epsilon/\Theta$ (compare eq. [1.108] and [U1]):

$$\Theta = \frac{\omega\varepsilon}{\ln\{1 + 2\omega\varepsilon/[2 - (\omega + 1 + R)\varepsilon]\}} \tag{2.90}$$

$$\omega = \sqrt{1 + R^2}$$

In practice, one can use these analytical solutions (2.88–2.90) for shell-and-tube heat exchangers with two tube-side passes and a large number of baffles. Four baffles or more may be regarded as sufficiently large in this context (see Fig. 2.19 and [G7]). The subdivision into single cells is only required for a small number of cells and for higher accuracy, especially at large NTU. Similar expressions can also be derived for apparatuses with a larger number of tube-side passes [F1, U1] (see also section 8 in this chapter). In order to complete a design, the required heat transfer coefficients for shell-and-tube heat exchangers with baffles may be calculated by the methods given in VDI-WA [V1] or in HEDH [H3].

Problems

2.9 Rewrite eq. (2.90) with $\epsilon_1 := \epsilon$ and $\epsilon_2 := R\epsilon$ in the form $\Theta(\epsilon_1, \epsilon_2)$. Would Θ change with interchanging the two streams?

2.10 Check with $N_1 := N$, $N_2 := RN$, whether it is necessary in eq. (2.88) to distinguish between the streams in the tubes (two passes) and on the shell side (one pass laterally mixed).

2.11 Derive an expression for the LMTD correction factor F (defined in eq. [1.110]) from eq. (2.29) (check your result against the diagram given by Gardner and Taborek in [G7] as Fig. 6).

2.12 Use the Gaddis–Vogelpohl method [G4] (see Figs. 2.15 and 2.16) to calculate two efficiencies of a shell-and-tube heat exchanger with two tube-side passes and four baffles (10 cells) with $N = 4.29$ and $R = 1$. Compare the result with the value obtained from eq. (2.88).

4 PLATE HEAT EXCHANGERS

4.1 Description

Plate heat exchangers of the type shown in Fig. 2.20 are presently offered by a large number of manufacturers as standard series production apparatus. They consist of a number of gasketed metal plates clamped between a stationary head and a follower plate by tie bolts. A wavy surface of a special design is stamped on the thin walled plates (see Fig. 2.21).

The plates are usually rectangular with circular ports at the four corners through which the two heat exchanging fluids may enter and leave. The gaskets are so arranged as to direct the two fluids through alternate flow channels formed by the space between the plates, as seen in Fig. 2.22. The corrugated surface pattern on the plates

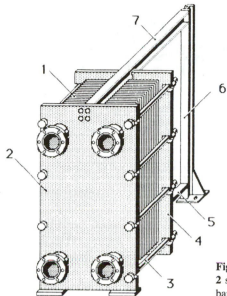

Figure 2.20 The plate heat exchanger. **1** plate pack, **2** stationary head, **3** tie bolt, **4** follower plate, **5** bottom bar, **6** end support, **7** top bar.

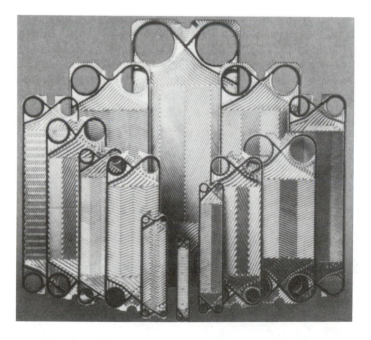

Figure 2.21 Plates with chevron-type corrugations (photo: Schmidt Comp.)

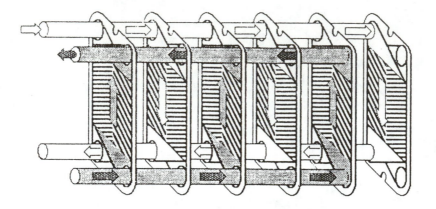

Figure 2.22 Parallel arrangement of plates.

subdivides the flow cross section into a number of interconnected parallel flow channels with multiple changes in direction as well as cross-sectional area.

The corrugations on adjacent plates crisscross, providing multiple points of contact and offering mechanical support against pressure difference across the plates. Using plates with only two or three of the four ports open, the path of the two streams through the pack can be so arranged that any number of channels may be connected in series (see Fig. 2.23).

A variety of connections is possible using a single element of construction built as an array in this type of heat exchanger, giving rise to a multitude of flow configurations. The main application area of these apparatuses is liquid-liquid heat transfer in the lower pressure range (usually below 1.6 MPa) because the construction and the large gaskets are unfavorable for high pressures in general. Due to the small interplate spacing and the high vorticity of the flow, one can reach high heat transfer

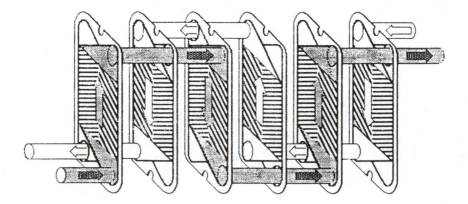

Figure 2.23 Series arrangement of plates.

coefficients. Ease of cleaning, simple adjustment to changed operating conditions by replacement or addition of plates, and the compactness and, correspondingly, a small liquid hold-up are usually cited as the most important advantages of plate heat exchangers.

Probably the most widely used configuration is the antiparallel arrangement of the flow paths of the two fluids through the plate back as in Fig. 2.22. The counterflow that results from this arrangement is usually desired and even needed in some applications. However, it is opposed in practice by a number of effects that tend to diminish the efficiency compared to that of an ideal counterflow heat exchanger.

4.2 A Peculiar End Effect

One of the causes of diminution in efficiency of the plate heat exchanger is the thermal end effect due to the two outermost streams of the plate pack taking up or losing heat only over one side, while all internal streams have both channel walls available for this purpose (see Fig. 2.24). Therefore, the number of transfer units N_1 and N_n for the two outermost streams are just half the value of all other N_j at uniform flow distribution if the number of channels n is even:

$$N_1 = \frac{1}{2}N_j \qquad j = 3, 5, 7, \ldots, n-1 \qquad (2.92)$$

$$N_n = \frac{1}{2}N_j \qquad j = 2, 4, 6, \ldots, n-2 \ (n \text{ even}) \qquad (2.92)$$

If n is odd, the arrangement has an adiabatic plane of symmetry in the middle of the central channel having the number $(n + 1)/2$. The problem may, therefore, be reduced to a pack with $(n + 1)/2$ channels having only one end channel, for which

$$N_1 = \frac{1}{2}N_j \qquad j = 3, 5, 7, \ldots, \frac{n+1}{2} - 1 \qquad (2.93)$$

a lower NTU in the end channels implies, of course, a smaller change in temperature of these streams.

As the streams are mutually coupled, the temperature distribution has to be found from a coupled system of linear differential equations, as shown in detail in [B1, B3], that can be written very compactly in vector notation (see also [D1, P3, Z1]):

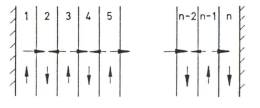

Figure 2.24 End effect in plate packs.

$$\boxed{\frac{d\Theta_j}{d\zeta} = [A]\Theta_j} \tag{2.94}$$

The coefficient matrix [A] has the form

$$[A] = \begin{pmatrix} -1 & 1 & & & & & \\ C & -2C & C & & & & \\ & 1 & -2 & 1 & & & \\ & & C & -2C & C & & \\ & & & & \cdots & & \\ & & & & 1 & -2 & 1 \\ & & & & & C & -C \end{pmatrix}$$

a tridiagonal matrix with n rows and columns. For $n = 2$, this leads to

$$\frac{d\Theta_j}{d\zeta} = \begin{pmatrix} -1 & 1 \\ C & -C \end{pmatrix} \Theta_j \tag{2.95}$$

and in written-out form:

$$\frac{d\Theta_1}{d\zeta} = -(\Theta_1 - \Theta_2) \tag{2.96}$$

$$\frac{d\Theta_2}{d\zeta} = C(\Theta_1 - \Theta_2) \tag{2.97}$$

These are the well-known equations for parallel and counterflow heat exchangers. Compare eqs. (1.59) and (1.60) with $\zeta = N_1Z$, $C\zeta = N_2Z$. The coefficient matrix in eq. (2.94) is written for even n. For odd n, only $(n + 1)/2$ rows and columns are needed, and the last row reads 0 2 -2 or 0 2C $-2C$, as can be seen from writing a balance for the central channel. Further, one has to consider that the "local" flow rate capacity ratio of individual streams for odd n is not equal to the total capacity ratio, unlike in the case of even n. Here we have $C_{total} = [(n + 1)/(n - 1)]C$. The general solution of the system of equations, (2.94), is

$$\Theta_j = \sum a_{jm} e^{r_m\zeta} \tag{2.98}$$

The complete solution is the sum of all terms from $m = 1$ to n. The eigenvalues r_m are the roots of the characteristic equation

$$\det|A - r_mE| = 0 \qquad E = \text{unit matrix} \tag{2.99}$$

For $n = 2$, the determinant is

$$\begin{vmatrix} -(1+r) & 1 \\ C & -(C+r) \end{vmatrix} = (1+r)(C+r) - C$$

and the roots are

$$r(r+1+C) = 0 \qquad r_1 = 0, \quad r_2 = -(1+C)$$

Therefore, the solutions is

$$\Theta_1 = a_{11} + a_{12}\, e^{-(1+C)\zeta} \tag{2.100}$$

$$\Theta_2 = a_{21} + a_{22}\, e^{-(1+C)\zeta} \tag{2.101}$$

The coefficients a_{jm} have to be determined from the differential equations and boundary conditions. For small numbers n, the solution can be obtained in closed form. The solution for $n = 4$ can be found in [B1, B3]. For larger numbers n, the analytical solution (eq. [2.98]) can still be used, but the characteristic equations have to be solved numerically. The limiting case of $C = 0$ is more easily treated, since the system of equations (2.94) then becomes completely decoupled. In practice, $C = 0$ can be realized by evaporation or condensation of a pure substance. Thus, the temperatures in all even-numbered channels might be kept constant: $\Theta_j = 0$ for $j = 2, 4, 6, \ldots$. The outlet temperatures on the other side, i.e., of the streams in channels 1, 3, 5, $\ldots$, can be easily calculated in this case:

$$\Theta_j'' = e^{-2N_j} \qquad j = 1, 3, 5, \ldots \tag{2.102}$$

With the phase change medium flowing in the even-numbered channels, there are two possibilities for the m streams on the other side: $m = n/2$ for even n and $m = (n + 1)/2$ for odd n. For even n, there is only one end channel with $N_1 = N_C$. For $j = 3, 5, 7, \ldots, (n - 1)$, $N_j = 2N_C$. Here N_C is based on the surface area A_P of one plate and flow rate $\dot{M}/m$:

$$N_C = \frac{kA_P}{\dot{M}c_p}\, m \tag{2.103}$$

There are $(n - 1)$ active plates, so that the number of transfer units for the whole apparatus becomes

$$N = \frac{kA_P}{\dot{M}c_p}\, (n - 1) \tag{2.104}$$

When the total number of channels is odd, the case of the constant temperature medium flowing through the odd-numbered channels is obviously trivial, as there is no end effect then. For the two remaining cases of n being even or odd, there are $(m - 1)$ or $(m - 2)$ inner streams and one or two end streams, respectively, and the exchanger outlet temperature results from mixing of the inner and outer streams. For even n, the efficiency is

$$\varepsilon = 1 - \frac{\exp\left(-\frac{nN/2}{n-1}\right) + ([n/2] - 1)\exp\left(-\frac{nN}{n-1}\right)}{n/2} \tag{2.105}$$

and for odd *n:*

$$\varepsilon = 1 - \frac{\exp\left(-\frac{(n+1)N/2}{n-1}\right) + ([(n + 1)/4] - 1)\exp\left(-\frac{(n+1)N}{n-1}\right)}{(n + 1)/4} \tag{2.106}$$

Substituting $n' = (n + 1)/2$ (symmetry for odd *n*) in eq. (2.106) yields eq. (2.105) with n' in place of *n*. From this, it follows for $C = 0$

$$\varepsilon(n') = \varepsilon\,(n)_{\text{even } n} = \varepsilon\left(\frac{n + 1}{2}\right)_{\text{odd } n} \tag{2.107}$$

Equation (2.105) can also be solved explicitly for N_C or for $N(\epsilon, n')$, respectively (problem!). With the basic relation $\epsilon = N\Theta$, one can also find $\Theta(\epsilon, n')$ or the LMTD correction factor $F(\epsilon, n')$:

$$F(\varepsilon, n')_{C=0} = \frac{n'}{2(n' - 1)} \frac{\ln(1 - \varepsilon)}{\ln\{[\sqrt{1 + n'(n' - 2)(1 - \varepsilon)} - 1]/(n' - 2)\}} \tag{2.108}$$

Figure 2.25 plots the loci of the values of the correction factor calculated from this equation against the number of channels n' with the efficiency as a parameter. Recall that $n' = n$ for even *n* and $n' = (n + 1)/2$ for odd *n*. It may appear somewhat surprising that the minimum in *F* for a given efficiency is not always found for $n' = 3$ (i.e., $n = 5$), for which we have the largest ratio of the number of end channels to total number channels on one side (i.e., 2/3). In other words, the consequence of the end effect is not always the greatest for $n' = 3$ as might be expected. At efficiencies greater than 0.7, the minimum shifts towards higher *n* for increasing efficiencies. This seemingly erroneous result may be explained as follows: In the definition of the correction factor, there are contained two effects—first, the thermal end effect, which decreases with increasing number of channels, and second a "geometrical end effect," expressing itself in the factor $n'/[2(n' - 1)]$ in eq. (2.108). The reciprocal of this factor is the number of active plates available for one individual stream. For $n' = 2$ (i.e., $n = 2$ or 3), this is just one plate per stream, while it becomes two for $n' \to \infty$. Thus, the first term in eq. (2.108), representing the geometrical end effect, goes monotonically from 1 to 0.5 as $n' \to \infty$, while the second term, representing the thermal end effect, is an increasing function of n' and tends to 2 as $n' \to \infty$. The correction factor, which is a product of these two terms, exhibits a minimum that varies with ϵ as in Fig. 2.25. With the phase change fluid flowing in even-numbered channels, comparison between two consecutive numbers *n* and $n + 1$ shows that the odd number is always less favorable (two end channels in place of one). If the connections of the two fluids are interchanged, then an odd number is

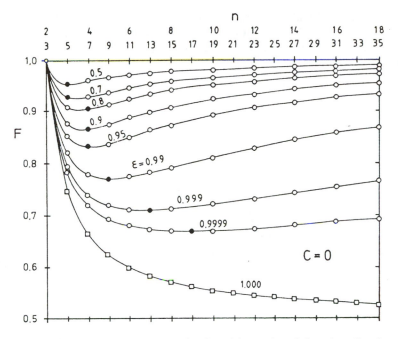

Figure 2.25 LMTD correction F as a function of the number of channels at $C = 0$.

obviously more favorable as the constant temperature fluid flows in the end channels (no end effect, $F = 1$).

Calculations for $C \neq 0$ were carried out analytically by Bassiouny [B1] and numerically, using a finite difference scheme, by Shah and Kandlikar [S9]. As far as the ranges of parameters overlap, these results agree very well with each other. The calculations for $C = -1$ (counterflow with equal capacity flow rates) show that the higher value of F is found for the odd number compared to the preceding or following even number of channels. The explicit solutions for $n = 5$ (three equations, $r_1 = r_2 = 0$, $r_3 = -5/3$) and $n = 4$ (4 equations, $r_1 = r_2 = 0$, $r_3 = \sqrt{2}$, $r_4 = -\sqrt{2}$) yield

$$F(n = 4, C = -1) = \frac{2}{3} + \frac{1}{3} \frac{1}{N/3 + (\sqrt{2}N/3)\coth(\sqrt{2}N/3)} \qquad (2.109)$$

$$F(n = 5, C = -1) = \frac{9}{10} + \frac{1}{10} \frac{1 - \exp(-5N/4)}{5N/4} \qquad (2.110)$$

The correction factors calculated from these equations have a value of unity for both cases for $N \to 0$ and as $N \to \infty$, drop to 2/3 or 9/10, respectively. The lower limit of the LMTD—correction for $C = 0$ is given by eq. (2.108) as $F_\infty = n/[2(n-1)]$ for even n and as $(n+1)/[2(n-1)]$ for odd n. The analytical results for $n = 4$ and those for $n = 6, 8, 10, 12$ with numerically calculated eigenvalues r_m lead to the same limit for $0 > C \geq -1$. With odd numbers n, however, different values of C result in different limits for F. The values calculated at $C = -1$ for $n = 5$ to 17 can be summarized by the formula

$$F_\infty(C = -1)_{n \text{ odd}} = \frac{3}{4}\frac{n+1}{n} \tag{2.111}$$

which, like that for even numbers, cannot claim general validity for arbitrary n.

Figure 2.26 shows the results of an experimental check [B1] of the end effect in an industrial plate-heat exchanger with $n = 4, 6, 12$ channels and $C = -1$. The correction factors calculated from measured data for packs of 4, 6, and 12 channels (i.e., 5, 7, and 13 plates) agree with the trend of the theoretical curves. The considerably lower absolute values, however, indicate that there are probably other reasons, apart from the end effect, for a deviation from ideal counterflow behavior. A maldistribution of flowrate in the parallel channels might be one of these reasons for lower efficiency. The counterflow effect that remains is, nevertheless, still considerable (esp. for 12 channels), as may be seen from the comparison with the curve for ideal crossflow (broken line). Calculations of temperatures, efficiencies, and correction factors for more complex flow configurations may be found in Bassiouny's thesis [B1], as well as in the earlier mentioned paper by Shah and Kandlikar [S9].

In the next section, some of the more important arrangements will be treated approximately for a large value of n, i.e., without taking end effects into account.

4.3 Series-Parallel Arrangements

If the flow rates of the two fluids are widely different, then the flow velocities and, consequently, the heat transfer coefficients can become very small for the stream with the lower flow rate, when the two streams are in simple counterflow as in Fig. 2.24. This may be avoided by connecting the flow channels of the smaller stream in series. Figures 2.27 and 2.28 show two of these arrangements.

Here one stream is directed in parallel to a set of alternate plate channels. The

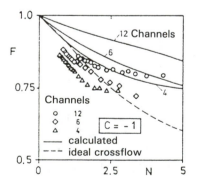

Figure 2.26 LMTD correction F vs. N at $C = -1$ for 4, 6, 12 channels. Theoretical results in comparison with experimental values (symbols).

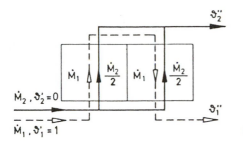

Figure 2.27 Parallel-series arrangement (1 × 2).

other stream flows through the remaining alternate plate channels in series. In the figures, the plate pack is shown divided into two and three equal segments, respectively. The series stream is alternately in co-current and counterflow to the stream connected in parallel. Such series-parallel arrangements may, of course, be realized with other types of heat exchangers also (e.g., double-pipe heat exchangers). The total surface area is divided into two or more parts. If the end effects and the effects of coupling of the various parts are neglected, each part may be treated as a separate parallel- or counterflow heat exchanger. Such an approximation is permissible for a sufficiently large number of parallel channels within each partial pack and for not-too-large NTUs.

The analysis is then very simple. For the 1 × 2 series-parallel arrangement of Fig. 2.27, one calculates first the outlet temperature ϑ_1^i of stream 1 (the intermediate temperature) from the first partial pack:

$$\frac{1 - \vartheta_1^i}{1 - 0} = \varepsilon_{\mathrm{p}}\left(\frac{N_1}{2}, 2R\right) \tag{2.112}$$

and then the outlet temperature from the second pack:

$$\frac{\vartheta_1^i - \vartheta_1''}{\vartheta_1^i - 0} = \varepsilon_{\mathrm{c}}\left(\frac{N_1}{2}, 2R\right) \tag{2.113}$$

where ε_{p} denotes the efficiency of a parallel, ε_{c} that of a counterflow heat exchanger. The arguments of these functions are $N_1/2$ and $2R$, because the whole stream 1 and only half of stream 2 flow through the partial pack with half the surface area. From the two eqs. (2.112) and (2.113), the efficiency of the 1 × 2 arrangement is found with $\epsilon = 1 - \vartheta_1''$.

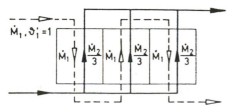

Figure 2.28 Parallel-series arrangement (1 × 3).

$$(1 - \varepsilon)_{1 \times 2} = \left[1 - \varepsilon_p\left(\frac{N_1}{2}, 2R\right)\right]\left[1 - \varepsilon_c\left(\frac{N_1}{2}, -2R\right)\right] \tag{2.114}$$

Here from eqs. (2.12) and (2.14), we have

$$1 - \varepsilon_p = \frac{e^{-(1+C)N} + C}{1 + C} \tag{2.115}$$

and

$$1 - \varepsilon_c = \frac{1 + C}{e^{+(1+C)N} + C} \qquad (C \neq -1)$$

$$1 - \varepsilon_c = \frac{1}{1 + N} \qquad (C = -1) \tag{2.116}$$

If these expressions are inserted with the arguments $N = N_1/2$ and $C = 2R$ or $C = -2R$, respectively, into eq. (2.114), explicit formula for the efficiency of the 1×2 series-parallel arrangement is obtained:

$$\varepsilon = 1 - \frac{1 - 2R \, \exp[-(1 + 2R)N_1/2] + 2R}{1 + 2R \, \exp[+(1 - 2R)N_1/2] - 2R} \qquad \left(R \neq \frac{1}{2}\right)$$

$$\varepsilon = 1 - \frac{1 + e^{-N_1}}{2 + N_1} \qquad \left(R = \frac{1}{2}\right) \tag{2.117}$$

Check: With $R = 0$, i.e., $\vartheta_2 = 0$, this again must result in $\epsilon = 1 - e^{-N_1}$!

Obviously, with this configuration, one can reach a maximum efficiency of $\epsilon_\infty = 1$ as $N_1 \to \infty$, given $R = 1/2$. For $N_1 = 10$, 15, and 20 and $R = 1/2$, 1/3, and 1/4, the efficiencies calculated numerically for $n \to \infty$ by Shah and Kandlikar [S9] for the 1×2 arrangement agree with those calculated by eq. (2.117). (Note: The subscripts 1,2 used in [S9] are the converse of our notation in Fig. 2.27).

According to these calculations, the end and coupling effects may be neglected in practice for the channels numbering more than 40 for the whole pack. At very large NTU and for a smaller number of channels, these effects are significant, however. In the case of the 1×3 series parallel arrangement, one can find the intermediate temperatures ϑ_1^{1i}, ϑ_1^{2i} by a procedure analogous to that for the 1×2 case:

$$\frac{\vartheta_1^{1i}}{1} = 1 - \varepsilon_c\left(\frac{N_1}{3}, -3R\right) \tag{2.118}$$

$$\frac{\vartheta_i^{2i}}{\vartheta_1^{1i}} = 1 - \varepsilon_p\left(\frac{N_1}{3}, 3R\right) \tag{2.119}$$

$$\frac{\vartheta_1''}{\vartheta_1^{2i}} = 1 - \varepsilon_c\left(\frac{N_1}{3}, -3R\right) \tag{2.120}$$

By multiplying these terms, the intermediate temperatures cancel, and one obtains a result that could easily be generalized for arbitrary 1 × m arrangements:

$$(1 - \varepsilon)_{1 \times 3} = \left[1 - \varepsilon_c \left(\frac{N_1}{3}, -3R \right) \right]^2 \left[1 - \varepsilon_p \left(\frac{N_1}{3}, 3R \right) \right]^1 \qquad (2.121)$$

Here it was assumed, as in Fig. 2.28, that two of the three passes are in counterflow and one in parallel flow. For the less favorable inverse case (two parallel, one counterflow), one has merely to interchange the exponents 2 and 1 in eq. (2.121). In a similar way, one can get equations for efficiency ε and normalized mean temperature difference $\Theta = \varepsilon/N$ for other plate heat exchanger configurations also. A detailed calculation accounting for the temperature variation in every single channel is required only for small numbers of plates and high NTU.

4.4 Pressure Drop and Heat Transfer

Because of the complex path of the fluid in the clearance between two corrugated plates (see Fig. 2.21), the conventional standard formulas for friction factors and heat transfer coefficients are not applicable here. As the details of the pressed patterns vary from manufacturer to manufacturer, one can hardly give generalized correlations. Nearly all the manufacturers today offer plates with chevron-type corrugations (as in Fig. 2.21) with the direction of the crests and troughs of the waves shifted by an angle φ from the longitudinal axis of the plate. With $\varphi = 0$, i.e., longitudinal waves, one would get parallel channels, separated from each other, which would have the lowest flow resistance. With $\varphi = 90°$, one would get transverse waves with infinitely large flow resistance (if the crests in the two plates are everywhere in contact). Both these extreme cases are mechanically less sturdy. Usually plates with either $\varphi \approx 30°$ or $\varphi \approx 70°$ are offered by the manufacturers.

Plate packs may be built up from equal plates, every second one turned around its surface normal by 180°, or alternately from dissimilar plates having different angles. The acute angle pattern ($\varphi = 30°$) yields channels with lower flow resistance. Such plates are called "soft" plates (S-plates). The stronger deviation from the longitudinal direction for the pattern with $\varphi = 70°$ results in correspondingly higher flow resistances. Plates of this type are called "hard" plates (H-plates). The pressure drops measured by Bassiouny [B1] using plates of one manufacturer (W. Schmidt GmbH and Co KG Bretten, Germany) with $\varphi = 71°$ (called Sigma-27 H) and $\varphi = 29°45'$ (Sigma-27 S) are plotted as friction factors ξ vs. the Reynolds number in Fig. 2.29.

As a characteristic length, the hydraulic diameter $d_h \approx 2b$ has been chosen. The gap width b is the distance between two planes touching the wave crests of a plate on either side minus the wall thickness. The flow velocity is defined as the volumetric rate $\dot{V}$ divided by $b \cdot B$. Here B is the width (or breadth) of the plate perpendicular to the mainflow direction measured between the gaskets. The "hard" pattern has about an eightfold higher flow resistance than the "soft" one. The friction factor for a channel built from one H- and one S-plate (H/S) is close to the arithmetic mean of the factors obtained with uniform channels built from H-plates or S-plates alone. The pressure drops were measured inside the plates, so that turnaround and friction losses in the distributors and collectors still have to be added to find the total pressure drop between the inlet and outlet headers of the apparatus.

Measurements of heat transfer for the same three channels (H, S, and H/S) are shown in Fig. 2.30 in terms of $Nu\,Pr^{-0.4}$ vs. Re from [B1]. In the range investigated, straight lines with slopes around 0.7 were found in this log-log plot, so that one can write

$$Nu = c\,Re^a\,Pr^{0.4} \tag{2.122}$$

with

φ	71°(H)	29°45′ (S)	(H/S)
a	0.69	0.72	0.70
c	0.274	0.094	0.184

in a range of validity of about $10^2 < Re < 10^4$, and $2 < Pr < 40$.

With water in the "hard" channels, for example, a pressure drop of 1 bar/m

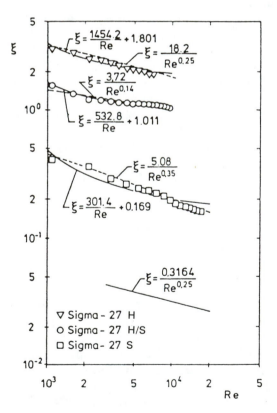

Figure 2.29 Friction factors $\xi(Re)$ for plate heat exchangers.

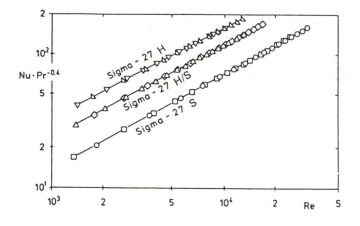

Figure 2.30 Heat transfer coefficients $Nu\ Pr^{-0.4}(Re)$ for plate heat exchangers.

results in flow velocities of 0.8 m/s and heat transfer coefficients of 23000 W/(m² K) ($k \approx 7500$ W/(m² K)). In the "soft" channels, however, we get about 1.7 m/s with $\alpha \approx 18000$ W/(m² K) ($k \approx 6300$ W/(m² K)). The overall heat transfer coefficients given in parentheses were calculated with a thermal conductivity of the wall of $\lambda_w \approx 15$ W/(K m) for stainless steel and a wall thickness of $s = 0.7$ mm ($1/k = 2/\alpha + s/\lambda_w$). Data for water are taken at an average temperature of 40 °C. The underlying dimensions are [B1] $d_h = 2b = 7.2$ mm, plate width $B = 260$ mm, plate length (measured between inlet and outlet ports) $L = 980$ mm, area $A_P = 0.25$ m².

With a correspondingly high pressure drop and, therefore, a high pumping power $\Delta p \dot{V}$, very high overall heat transfer coefficients can be reached with these apparatuses. At $\Delta p/\Delta L \approx 0.1$ bar/m, one typically obtains flow velocities of $u = 0.20$ m/s (for H), 0.46 m/s (for S) with $\alpha_H \approx 9000$ W/(m² K) ($k \approx 3700$ W/(m² K)) and $\alpha_S \approx 6700$ W/(m² K) ($k \approx 2900$ W/(m² K)), respectively. From these values and eq. (2.122), heat transfer coefficients in plate heat exchangers may also be estimated for fluids other than water. Similar correlations (as in Figs. 2.29 and 2.30) for pressure drop and heat transfer in plate heat exchangers can be found in HEDH [H3] 3.7.3 (supplement 1989) for chevron-type plates with angles of $\varphi = 30°$, $40°$, and $60°$ (there the angle $\beta = 2\varphi$ is given [60°, 80°, 120°]).

5 SPIRAL PLATE HEAT EXCHANGERS

5.1 Description

Figure 2.31 shows a spiral plate heat exchanger with counterflow of the two fluids inside the double spiral formed by winding two metal sheets around a cylindrical core. Steel bolts are provided at regular intervals to serve as spacers to maintain a constant distance between the plates as well as to stiffen the plates against fluid

Figure 2.31 Spiral-plate heat exchanger (photo and drawing: Schmidt Comp.).

pressure. Each spiral channel is sealed by welding on one edge, the other side being accessible for inspection and cleaning. The two spiral channels are, thus, open on opposite sides and are closed with a lid-plate and a flat gasket during operation. The welded edges of the spiral channels have to be accurately machined to provide a leakproof closure against the flat gasket. Manufacture of spiral plate exchangers is more labor intensive, and, hence, they are more expensive compared with plate exchangers built from pressed plates. Since there exists only one continuous flow channel for each fluid, potential fouling deposits are swept away more easily and positively. The "hot end" of the counterflow arrangement may be placed at the core of the spiral exchanger, shielding it more or less completely from the surroundings. These exchangers are, therefore, often used for energy recovery, i.e., preheating of process fluids by hot product streams which have to be cooled. The spiral plate exchangers are especially advantageous in handling solid suspensions. In a slightly modified design, they can be used as compact reflux condensers directly on top of a distillation column. The largest units offered have surface areas of up to 400 m² [M1].

5.2 Temperature Profile and Mean Temperature Difference

The calculation of the temperature profile in a spiral plate heat exchanger is more involved than in an ideal counterflow exchanger. This is due to the fact that each cold stream is separated from an adjacent cold stream by one turn of the spiral and both receive heat from the hot fluid flowing in between. Analytical solutions of the problem have been given in 1983 by Cieslinski and Bes [C3] in terms of series expansions of Hermite polynomials and in 1988 by Bes [B3a] in terms of series expansions of exponential functions. These analytical solutions are, however, hardly suitable for practical application, as the evaluation of the series requires tedious calculations. The authors have, therefore, provided charts of $\Delta\epsilon$, the efficiency reduction with respect to ideal counterflow as a function of NTU, capacity rate ratio C, and number of turns [C3], as well as charts of ϵ_1 vs. ϵ_2 with Θ and N_1 as a parameter for a few numbers of half turns [B3a]. Numerical calculations have been done for up to 16 turns and NTU up to 10 by Chowdhury et al. [C1]. These have been done for the three types of double spirals shown schematically in Fig. 2.32, differing only in the relative posi-

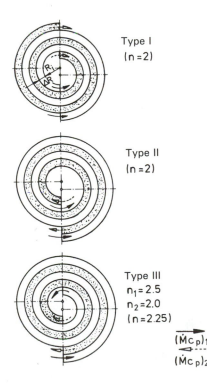

Type I
(n = 2)

Type II
(n = 2)

Type III
$n_1 = 2.5$
$n_2 = 2.0$
(n = 2.25)

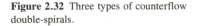

Figure 2.32 Three types of counterflow double-spirals.

tions of inlet and outlet at the inner and outer ends of the spirals, respectively. As an example of these results, Figs. 2.33 and 2.34 show the temperature profile vs. the flow length for types I and II of Fig. 2.32 for a high NTU ($N = 10$) and a small number of turns ($n = 4$), with counterflow at equal capacities ($C = -1$). At such a small number of turns, the end effects (heat flux only to one side of the channel in the innermost and outermost half turns), as well as the general deviation from the linear profile of an ideal counterflow exchanger, are evident. The reduction in efficiency, $\Delta\epsilon$, can be directly read from these graphs.

Of the several possible ways to present the profusion of numerical results ($\Delta\epsilon$ depends on N, C, n, and the position of inlets, Type I, II, III, . . .), a compact form was obtained with the plot of the LMTD correction factor F for constant C versus the ratio N/n, i.e., the NTU per turn. Figures 2.35 and 2.36 show this plot of the numerical results for $C = -1$. Only for very small numbers of turns, one may find additional influences of type and number of turns. In the ranges of practical interest, it is obviously possible to reduce the number of parameters from four (N, C, n, type) to two (N/n, C). The formal similarity between the F vs. (N/n) curve and the well-known variation of fin efficiency vs. fin height for plane fins (see Fig. 3.23) inspired the choice of the function $\tanh(x)/x$, describing this fin efficiency, as an empirical fitting curve for the numerical data:

$$F \approx \frac{n}{N} \tanh \frac{N}{n} \qquad (C = -1) \qquad (2.123)$$

It was found that this simple analytical function could represent the numerical results strikingly well without any additional parameter. The analytical results of Cieslinski and Bes [C3] as well as Bes [B3a] are also in agreement with this, except for small differences at low numbers of turns. The successful approximation of the

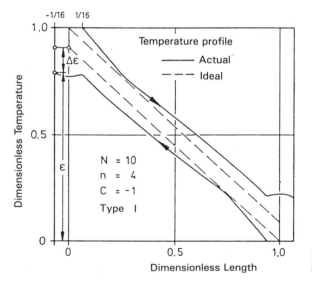

Figure 2.33 Temperature variation in the spiral-plate heat exchanger (type I).

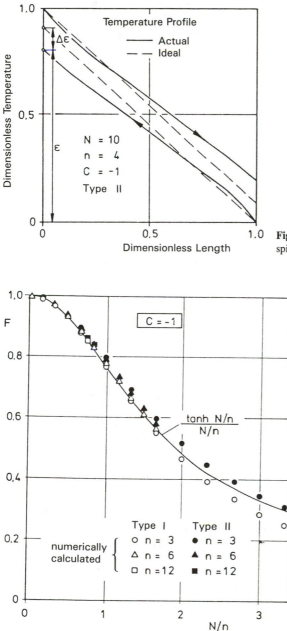

Figure 2.34 Temperature variation in the spiral-plate heat exchanger (type II).

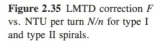

Figure 2.35 LMTD correction F vs. NTU per turn N/n for type I and type II spirals.

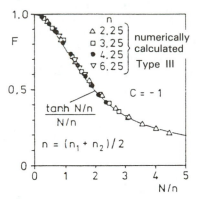

Figure 2.36 LMTD correction F vs. NTU per turn N/n for type III spirals.

numerical results by such a simple function as eq. (2.123) encourages one to hope that there may exist a simpler physical model for the thermal behavior of the spiral plate heat exchanger, which would lead analytically to the functional form of eq. (2.123) that had been first introduced purely as an empirical curve fit.

Trying out several configurations of cascades (see section 3.1), it is found that a countercurrent cascade of n parallel flow exchangers (see Fig. 2.37) is exactly what we are looking for!

From eq. (2.83) for a countercurrent cascade of n equal elements at $C = -1$ $(R = 1)$, the following relation is derived:

$$\frac{\varepsilon}{1-\varepsilon} = n \frac{\varepsilon_i}{1-\varepsilon_i} \tag{2.124}$$

Inserting for ε_i the efficiency of a parallel flow cell from eq. (2.115) but with $C = |C| = 1$ for parallel flow, we obtain

$$\frac{\varepsilon}{1-\varepsilon} = n \frac{1-e^{-2N/n}}{1+e^{-2N/n}} \tag{2.125}$$

With $F\Theta_{LM} = \varepsilon/N$ and $\Theta_{LM} = 1 - \varepsilon$ (for $C = -1$), one eventually finds that the LMTD correction factor for a countercurrent cascade of n equal cocurrent (parallel flow) elements is

$$F = \frac{n}{N} \frac{e^{+N/n} - e^{-N/n}}{e^{+N/n} + e^{-N/n}} \tag{2.126}$$

or

$$\boxed{F = \frac{n}{N} \tanh \frac{N}{n}} \qquad (C = -1) \tag{2.127}$$

At first glance, there appears to be no similarity between the cascade model from Fig. 2.37 and the spiral plate heat exchanger. The equivalence may be seen if the double

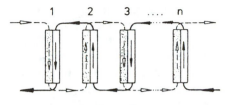

Figure 2.37 Countercurrent cascade of n parallel flow exchangers.

spiral is cut into two halves (see Fig. 2.38a). The cut halves may be represented by two plate packs interconnected as shown in Fig. 2.38b, where the half-channels are symbolized by equal rectangular shapes. The half-channels are numbered in the flow direction on the hot side as 0 to 5 and against the flow direction on the cold side as 0' to 5'. From any given internal channel on the hot side (1, 2, 3, or 4), heat is transferred to two adjacent cold streams with different temperatures. If we *replace the heat fluxes from stream 1 to the two cold streams 0' and 2' by a single heat flux to the stream 1'* whose temperature is half-way between those of the streams 0' and 2', we obtain the simplified configuration c in Fig. 2.38 which is the countercurrent cascade of parallel flow elements.

Since the simplified model does not account for the end effects and for the different positions of inlet and outlet of the three types investigated, it can only be

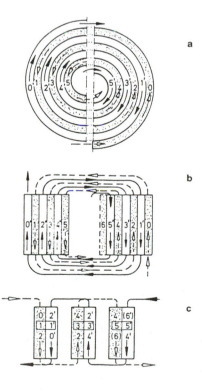

Figure 2.38 Development of a simpler model for the spiral-plate heat exchanger.

valid for sufficiently large numbers of turns. This is demonstrated in Fig. 2.39 where the ratio of $F_{(numerical)}$ to $F_{(model)}$ (from eq. [2.127]) is plotted against the number of turns n for $C = -1$ and the largest numerically investigated value of NTU ($N = 10$).

The deviation of the numerical results from the model predictions lies between 0 and $+5\%$ for $n > 6$. For smaller numbers of turns, the end effects become increasingly important. For $n = 1$, type II becomes an ideal counterflow exchanger ($F = 1$), while type I becomes a cocurrent cascade of two counterflow elements ($F = 1/[1 + (N/2)^2]$). Type II is always more favorable than type I. As is to be expected, type III lies in between. For $n > 10$, the differences are practically negligible. The model can be applied to other capacity rate ratios too. For $C \neq -1$, one obtains the following expressions for ϵ_i and F:

$$\boxed{F\left(\frac{N}{n}, C\right) = \frac{1}{(1+C)N/n} \ln\left(1 + \frac{1+C}{(1/\varepsilon_i) - 1}\right)} \qquad (C \neq -1) \qquad (2.128)$$

and

$$\varepsilon_i = \frac{1 - e^{-(1+|C|)N/n}}{1 + |C|} \qquad (2.129)$$

As before, the correction factors calculated from these equations (for the countercurrent cascade of n parallel flow elements) agree very well with the numerical results for spiral plate heat exchangers, as may be seen from Fig. 2.40. So the countercurrent cascade of n parallel flow elements is adequate as a model for the design of spiral plate heat exchangers, except for the cases of small numbers of turns and high NTU, which are, however, less important in practice.

High efficiencies at large NTU can only be reached with a large number of turns. For $C = -1$, one obtains from the cascade model a maximum efficiency of

$$\varepsilon_\infty(C = -1) = \frac{n}{n+1}$$

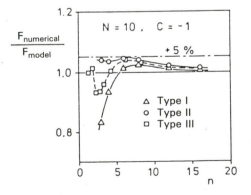

Figure 2.39 Comparison of numerically calculated LMTD correction factors with those obtained from the cascade model vs. number of turns.

More than 19 turns are therefore required to reach efficiencies above 95%. Recently Bes and Roetzel [B3b] have shown that the maximum efficiency for a type II spiral (as calculated from an exact analytical solution) is not reached asymptotically for $N \to \infty$, but at a finite, large value of NTU (usually greater than 15), with decreasing efficiencies for increasing NTU above this maximum (as in the case of cross flow, both sides laterally mixed). They also derived a new analytical approximation for F, which may be written in a slightly rearranged form as

$$F = z \ln[1 + 1/(z + x)] \tag{2.130}$$

with

$$x = y^2/(1 + 2y), \quad y = r_i/(2n\Delta r), \text{ and } z = (1 + 2y)/[(N/n)^2 R].$$

In addition to N/n and R, this formula contains the parameter y, i.e., the core radius r_i divided by the width of the double spiral pack of $2n$ channels.

Problem

Determine the limiting curves $F(N/n)$ for $y \to \infty$ (large core diameter) and $y \to 0$ (small core diameter) and the maximum efficiency ϵ_{max} from eq. (2.130) and compare the results with those from the cascade model (eqs. [2.127–2.129]!).

5.3 Pressure Drop and Heat Transfer

As the channels of a spiral plate exchanger are of a rectangular cross-section, pressure drop and heat transfer can be calculated in principle from the standard formulas for parallel plate ducts (annuli with $K \to 1$). The pressure drop will be increased because of the plate spacers and, at intermediate Reynolds numbers, also the heat transfer. The effect of the plate curvature is similar to that of the spacers. However, the curvature effect should be of minor importance as the core diameters are usually quite large compared to the gap width. Measurements [C2] with spiral plate exchang-

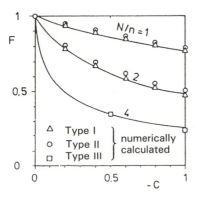

Figure 2.40 LMTD correction F vs. capacity flow rate ration C.

ers have shown that the friction factor can be calculated as the sum of a value for laminar flow in a parallel plates duct (from Eq. [1.116] with $K \to 1$ and $\varphi = 1.5$) and another term, $\text{const} \cdot Re^{-0.1}$, probably resulting from flow separation at the spacer bolts.

$$\xi = 1.5 \frac{64}{Re} + c\, Re^{-0.1} \tag{2.131}$$

The apparatus used in the investigations had a cross section of 5×300 mm^2, number of turns $n = 8.5$, core diameter of 250 mm, outer diameter of 495 mm and 5×5 mm cylindrical bolts in a rectangular in-line arrangement of 61×50 mm. For data in the range $4 \cdot 10^2 < Re < 3 \cdot 10^4$, the constant was found to be $c = 0.2$. Heat transfer data in the same range with water as the medium could be correlated by

$$Nu = 0.04\, Re^{0.74} Pr^{0.4} \tag{2.132}$$

In the upper range of Reynolds numbers, there is fair agreement between eq. (2.132) and the standard equation for turbulent flow in a parallel plate duct (eq. [1.140] with $d_{\mathrm{h}}/L \approx 0$, $K \to 1$ and hence, $f_a = 0.86$) as can be seen from the following table:

Re	2 300	5 000	10 000	30 000
$Nu_{(1.140)}$	13.3	34.7	68.3	182
$Nu_{(2.132)}$	26.8	47.6	79.5	179

In the intermediate and lower range of Reynolds numbers, the values are significantly higher due to vortex formation at the spacer bolts, and, to a lower degree, due to the plate curvature.

6 COUPLING OF TWO HEAT EXCHANGERS BY A CIRCULATING HEAT CARRIER

For reasons of safety or spatial constraints, it may be necessary to carry out heat transfer between two streams not directly in one apparatus, but in two separate heat exchangers coupled by a circulating heat carrier. Figure 2.41 shows such a connection schematically using symbols convenient for this case and the conventional definitions of N_1, N_2, R_1, R_2. The efficiency of the whole arrangement can be expressed in terms of the efficiencies ϵ_{11}, ϵ_{22}, and the capacity rate ratios R_1 and R_2 (the first subscript in ϵ_{ij} stands for the stream, the second for the apparatus):

$$\frac{\vartheta_1' - \vartheta_1''}{\vartheta_1' - \vartheta_{\mathrm{S1}}'} = \varepsilon_{11}(N_1, R_1) \tag{2.133}$$

$$\frac{\vartheta_2'' - \vartheta_2'}{\vartheta_{\mathrm{S2}}' - \vartheta_2'} = \varepsilon_{22}(N_2, R_2) \tag{2.134}$$

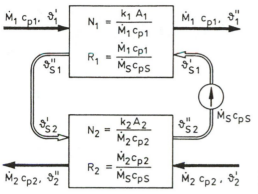
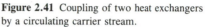

Figure 2.41 Coupling of two heat exchangers by a circulating carrier stream.

Apart from these two equations, which follow immediately on applying the definitions of ϵ_i to the two exchangers, one can write down the steady state balance for each of the apparatuses:

$$R_1(\vartheta_1' - \vartheta_1'') = \vartheta_{S1}'' - \vartheta_{S1}' \tag{2.135}$$

$$R_2(\vartheta_2'' - \vartheta_2') = \vartheta_{S2}' - \vartheta_{S2}'' \tag{2.136}$$

If the pumping power added to and the heat lost from the circulating stream are neglected, then we also have

$$\vartheta_{S1}' = \vartheta_{S2}'' = \vartheta_S' \tag{2.137}$$

and

$$\vartheta_{S2}' = \vartheta_{S1}'' = \vartheta_S'' \tag{2.138}$$

In order to eliminate the temperatures ϑ_S' and ϑ_S'', eqs. (2.133) and (2.134) are solved for these intermediate temperatures of the circulating stream:

$$\vartheta_S' = \vartheta_1' - \frac{\vartheta_1' - \vartheta_1''}{\varepsilon_{11}} \tag{2.139}$$

$$\vartheta_S'' = \vartheta_2' + \frac{\vartheta_2'' - \vartheta_2'}{\varepsilon_{22}} \tag{2.140}$$

The differences on the right hand side of eqs. (2.135) and (2.136) now become

$$\vartheta_S'' - \vartheta_S' = -(\vartheta_1' - \vartheta_2') + \frac{\vartheta_1' - \vartheta_1''}{\varepsilon_{11}} + \frac{\vartheta_2'' - \vartheta_2'}{\varepsilon_{22}} \tag{2.141}$$

The intermediate temperatures can be thus eliminated. Replacing $\vartheta_2'' - \vartheta_2'$ by $(R_1/R_2)\cdot(\vartheta_1' - \vartheta_1'')$ (from a total balance or the combination of the two partial balances) and dividing by $(\vartheta_1' - \vartheta_1'')$, one obtains

$$\boxed{\frac{1}{\varepsilon_1} = \frac{1}{\varepsilon_{11}(N_1, R_1)} + \frac{R_1}{R_2 \varepsilon_{22}(N_2, R_2)} - R_1} \tag{2.142}$$

Here ε_1 is defined as the change in temperature of stream 1 related to the maximum temperature difference of the whole arrangement:

$$\varepsilon_1 = \frac{\vartheta_1' - \vartheta_1''}{\vartheta_1' - \vartheta_2'} \tag{2.143}$$

Equation (2.142) may now be discussed with respect to the maximum efficiencies that can be reached with such an arrangement. We have to keep in mind that the functions $\varepsilon_{11}(N_1, R_1)$ and $\varepsilon_{22}(N_2, R_2)$ are the efficiencies of stream 1 in exchanger 1 and of stream 2 in exchanger 2, respectively. Then, $R_1 \varepsilon_{11}$ is the efficiency of the circulating stream S in exchanger 1, which can be obtained from the same function by renaming the arguments $(N_1 R_1 := N_1, 1/R_1 := R_1)$

$$R_1 \varepsilon_{11} = \varepsilon_{S,1} = \varepsilon_{11}\left(N_1 R_1, \frac{1}{R_1}\right) \tag{2.144}$$

and

$$R_2 \varepsilon_{22} = \varepsilon_{S,2} = \varepsilon_{22}\left(N_2 R_2, \frac{1}{R_2}\right) \tag{2.145}$$

Three cases can be distinguished here:

---

1. Both $R_i > 1$ (weak circulating flow)

$$\frac{1}{\varepsilon_1} = R_1\left(\frac{1}{\varepsilon_{11}(N_1 R_1, 1/R_1)} + \frac{1}{\varepsilon_{22}(N_2 R_2, 1/R_2)} - 1\right)$$

$$\implies \quad \varepsilon_{1\,max} = \frac{1}{R_1} < 1 \tag{2.146}$$

---

2. Both $R_i < 1$ (strong circulating flow)

$$\implies \quad \varepsilon_{1\,max} = \frac{1}{1 + R_1[(1/R_2) - 1]} < 1 \tag{2.147}$$

---

3. $R_1 \leq 1$ and $R_2 \geq 1$ $\qquad [(\dot{M}c_p)_1 \leq (\dot{M}c_p)_S \leq (\dot{M}c_p)_2]$

$$\frac{1}{\varepsilon_1} = \frac{1}{\varepsilon_{11}(N_1, R_1)} + R_1\left(\frac{1}{\varepsilon_{22}(N_2 R_2, 1/R_2)} - 1\right) \tag{2.148}$$

$$\implies \quad \varepsilon_{1\,max} = 1$$

Here we have assumed that the individual exchangers can, in principle, reach efficiencies of unity, i.e., that they are ideal counterflow or crossflow exchangers, or countercurrent cascades. We have further assumed that $(\dot{M}c_p)_1 \leq (\dot{M}c_p)_2$, i.e., $R_1 \leq R_2$ will be always valid, which does not restrict the generality of these statements. (If $R_1 > R_2$, one can simply regard ϵ_2 in place of ϵ_1 or rename the streams.) High total efficiencies can obviously be reached only if the circulating flow is chosen so that its flow capacity lies in between those of the two other streams.

The optimal flow rate of the circulating stream, as recently also shown by Roetzel [R4], is found from:

$$(\dot{M}_s c_{pS})_{\text{opt}} = \frac{k_1 A_1 + k_2 A_2}{N_1 + N_2} \tag{2.149}$$

if both apparatuses are counterflow heat exchangers.

Figures 2.42–2.44 show the efficiency of ϵ vs. $1/R_1$, i.e., the ratio of the circulating flow capacity to that of stream 1 for $R_1/R_2 = 0$, 0.5, and 1, respectively. The parameter of the curves is always N_1, with $N_2 = N_1$ for the dotted lines and $N_2 R_2 = N_1 R_1$ for the full lines. At low circulating flows $(Mc_p)_S$, the efficiency increases linearly with the circulating flow rate to fall again after passing a more or less sharp maximum (for $N > 1$ and $R_1/R_2 > 0$). In Fig. 2.43, one can clearly recognize the range $0.5 \leq R_1 \leq 1$, in which efficiencies up to unity are theoretically possible. At $R_1 = R_2$, this range shrinks to the point $R_1 = R_2 = 1$, and the circulating flow in this case has to be chosen very accurately if efficiencies above 80% are to be reached. The individual exchangers have been regarded as ideal counterflow ones here.

In the special case $R_1 = R_2 = 1$, with $\epsilon_{11}(N_1, 1) = N_1/(1 + N_1)$ and $\epsilon_{22}(N_2, 1) = N_2/(1 + N_2)$, it follows from eq. (2.142) that

$$\frac{1}{\epsilon} = \frac{1}{N_1} + \frac{1}{N_2} + 1 \tag{2.150}$$

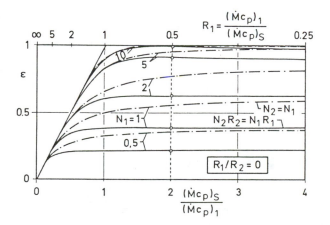

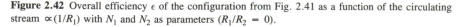

Figure 2.42 Overall efficiency ϵ of the configuration from Fig. 2.41 as a function of the circulating stream $\propto (1/R_1)$ with N_1 and N_2 as parameters ($R_1/R_2 = 0$).

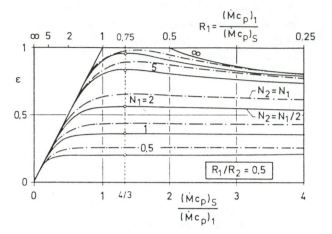

Figure 2.43 Overall efficiency ϵ of the configuration from Fig. 2.41 as a function of the circulating stream $\propto (1/R_1)$ with N_1 and N_2 as parameters ($R_1/R_2 = 0.5$).

If we further assume that the two exchangers are equal, $k_1 A_1 = k_2 A_2 = kA$, then we have

$$\frac{1}{\varepsilon} = 1 + \frac{2}{N} \tag{2.151}$$

or

$$N = \frac{2\varepsilon}{1 - \varepsilon} \tag{2.152}$$

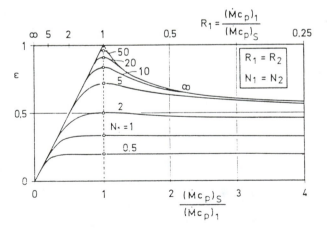

Figure 2.44 Overall efficiency ϵ of the configuration from Fig. 2.41 as a function of the circulating stream $\propto (1/R_1)$ with N_1 and N_2 as parameters ($R_1/R_2 = 1.0$).

With direct heat transfer in a single counterflow exchanger at equal flow capacities, we have

$$N_{\text{ideal}} = \frac{\varepsilon}{1 - \varepsilon} \tag{2.153}$$

Since N is expressed with the surface area A of one of the two exchangers, one would require a fourfold NTU with indirect transfer for the same efficiency in both cases. This does not necessarily imply a fourfold surface area. For gas-to-gas heat transfer in one apparatus, the overall heat transfer coefficient is roughly half the gas side heat transfer coefficient α_g:

$$k_{\text{g.g}} \approx \frac{\alpha_g}{2} \tag{2.154}$$

If a liquid is chosen as a circulating stream with much better heat transfer coefficient $\alpha_l \gg \alpha_g$, one can write approximately

$$k_{\text{g,l}} \approx \alpha_g \tag{2.155}$$

Then the surface area would have to be just twice as large for indirect heat transfer at the same α_g, same $\dot{M}c_p$, and the same heat rate $\dot{Q}$, compared to direct transfer in one exchanger. So one will choose an intermediate circuit only in those cases where direct transfer is forbidden by inevitable reasons.

7 REGENERATORS

7.1 Description

The coupling of two heat exchangers by a third heat carrier that circulates between the two has been described in the previous section. Similarly, the so-called regenerators also use a third medium as an intermediate store for the energy to be transmitted from the hot to the cold stream. In a regenerator, however, this intermediate storage medium is a solid matrix, which is heated and cooled (or charged and discharged) by the two fluid streams during alternate periods. The solid matrix may be built as a slowly rotating cylinder or disc as shown in Fig. 2.45.

This type of *regenerator with a rotating storage mass* has come to be known as the Ljungström air preheater [H2]. One can immediately recognize the similarity to the scheme shown in Fig. 2.41. The circulating heat carrier—here the slowly rotating disc—is in crossflow, however, to the fluids 1 and 2 in this case.

Figure 2.46 shows a schematic representation of this process from which not only the similarity, but also the differences to the problem treated previously (see Fig.

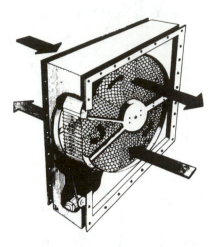

Figure 2.45 The rotating regenerator—the Ljungström air pre-heater.

2.41) may be clearly seen. In principle we have here a cross-counterflow connection. The older type of *regenerator with fixed storage masses,* of which the "classic" Cowper's towers for blast furnaces is an example, is shown schematically in Fig. 2.47. The hot and the cold gas streams alternate between the two storage beds periodically. Here it is not so easy to recognize that we have to deal with a countercurrent connection of two heat exchangers coupled in crossflow again. The role of the transverse coordinate y (corresponding to the flow direction of the coupling medium $\dot{M}_S$ in Fig. 2.46) is here taken over by the elapsed time after each switching of streams.

7.2 The "Short" Regenerator

To analyze the transfer performance of regenerators, we start, as for crossflow in section 2.1, with the relatively simple case of crossflow over one row of tubes or crossflow, one side laterally mixed. In Fig. 2.46, one can easily imagine this case as a disc-like section of the matrix (cut along the broken line), containing only a short slice of the rotating mass, corresponding to one of the circulating arrows in the scheme of Fig. 2.46.

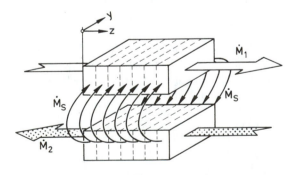

Figure 2.46 Scheme of the regenerator with circulating storage mass.

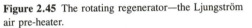

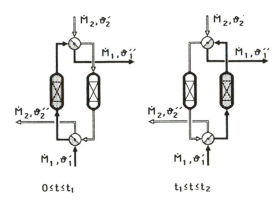

Figure 2.47 Scheme of the regenerator with fixed storage masses.

Figure 2.48*a*, *b*, shows the corresponding case of a "short" regenerator with fixed masses. The temperature of the solid does not depend on the longitudinal (or counterflow) coordinate z, but only on the transverse coordinate y (see Fig. 2.46) or on the time (see Fig. 2.48*a*, *b*). The energy balance for an element of the storage mass is

$$-\left(M_S c_{pS} \frac{dz}{L_z}\right)_i \frac{d\vartheta_S}{dt} = \left(kA \frac{dz}{L_z}\right)_i (\vartheta_S - \vartheta_i) \qquad (i = 1, 2) \qquad (2.156)$$

For the steadily rotating mass, we put $dt = (M_S/\dot{M}_S)\, dy/L_y$, L_z, and L_y being the total lengths in the two directions. A corresponding balance for the gas-side (stream 1 or 2) is

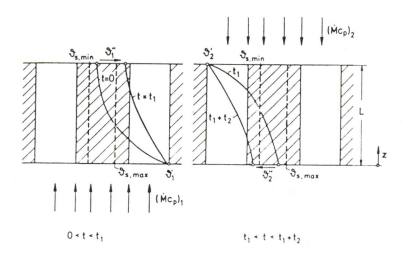

Figure 2.48 The "short" regenerator.

$$(\dot{M}c_{\mathrm{p}})_i \frac{\partial \vartheta_i}{\partial z}\, \mathrm{d}z = \left(kA\frac{\mathrm{d}z}{L_z}\right)_i (\vartheta_S - \vartheta_i) \tag{2.157}$$

if the storage capacity of the gas is neglected in comparison to that of the solid ("quasi steady state approximation"). In dimensionless form, we can write in the same fashion as for cross flow, one side laterally mixed (see eqs. [2.16 and 2.17]) ($i = 1, 2$):

$$-\frac{\mathrm{d}\vartheta_S}{\mathrm{d}\tau_i} = \vartheta_S - \bar{\vartheta}_i \tag{2.158}$$

$$\frac{\partial \vartheta_i}{\partial \zeta_i} = \vartheta_S - \vartheta_i \tag{2.159}$$

The dimensionless time—or length—coordinates vary between 0 and a number of transfer units $N_{\mathrm{S}i}$ or N_i, respectively. Since the differential equations are identical with those for crossflow, one side laterally mixed, one can use the solution developed in section 2.1. The mixed stream had the subscript 1 there and corresponds to the storage stream S here. From eq. (2.27), we can calculate the outlet temperature of the storage stream from exchanger 1 (or the final temperature of the storage mass after the "hot blowing period" t_1) ("S" := "1", "1" := "2"):

$$\varepsilon_{\mathrm{S1}} \equiv \frac{\vartheta_{\mathrm{S, max}} - \vartheta_{\mathrm{S, min}}}{\vartheta_1' - \vartheta_{\mathrm{S, min}}} = 1 - \exp[-R_1(1 - e^{-N_1})] \tag{2.160}$$

We take $N_{\mathrm{S1}} = k_1 A_1 t_1/(M_{\mathrm{S}}c_{\mathrm{pS}})_1$ or $k_1 A_1/(\dot{M}_{\mathrm{S}}c_{\mathrm{pS}})_1$ (dimensionless hot blowing period or NTU of the storage mass), and $N_1 = k_1 A_1/(\dot{M}c_{\mathrm{p}})_1$, as well as $R_1 = N_{\mathrm{S1}}/N_1 = \dot{M}_1 c_{\mathrm{p1}} t_1/(M_{\mathrm{S}}c_{\mathrm{pS}})_1$. Similarly, for the cooling of the storage mass in exchanger 2 or the "cold blowing period" t_2, it follows that

$$\varepsilon_{\mathrm{S2}} \equiv \frac{\vartheta_{\mathrm{S, max}} - \vartheta_{\mathrm{S, min}}}{\vartheta_{\mathrm{S, max}} - \vartheta_2'} = 1 - \exp[-R_2(1 - e^{-N_2})] \tag{2.161}$$

with N_{S2}, N_2, and R_2 defined as for N_{S1}, N_1, and R_1 above. With $\epsilon_{11} = (N_1/N_{\mathrm{S1}})\epsilon_{\mathrm{S1}} = \epsilon_{\mathrm{S1}}/R_1$ and ϵ_{22} similarly defined, the outlet temperatures of the gas streams can also be determined:

$$\varepsilon_{11} \equiv \frac{\vartheta_1' - \vartheta_1''}{\vartheta_1' - \vartheta_{\mathrm{S, min}}} = \frac{1 - \exp[-R_1(1 - e^{-N_1})]}{R_1} \tag{2.162}$$

$$\varepsilon_{22} \equiv \frac{\vartheta_2'' - \vartheta_2'}{\vartheta_{\mathrm{S, max}} - \vartheta_2'} = \frac{1 - \exp[-R_2(1 - e^{-N_2})]}{R_2} \tag{2.163}$$

From these four equations (2.160–163) the four temperatures ϑ_1'', ϑ_2'', $\vartheta_{\mathrm{Smin}}$, and $\vartheta_{\mathrm{Smax}}$, as well as the overall efficiency $\epsilon_1 = (\vartheta_1' - \vartheta_1'')/(\vartheta_1' - \vartheta_2')$ can be found. The solution follows the same path as in section 2.6 and, as expected, leads to the same result (see eq. [2.142]):

$$\frac{1}{\varepsilon_1} = \frac{1}{\varepsilon_{11}(N_1, R_1)} + \frac{R_1}{R_2 \varepsilon_{22}(N_2, R_2)} - R_1 \tag{2.164}$$

For the frequently encountered case of equal capacities $R_1 = R_2$ and equal exchangers $N_1 = N_2$, eq. (2.164), along with eqs. (2.162) and (2.163), yields, after some rearrangement,

$$\varepsilon = \frac{N}{N_S} \tanh\left(\frac{N_S}{2} \frac{1 - e^{-N}}{N}\right) \tag{2.165}$$

From eq. (2.165), the following limits may be derived:

$$\lim_{N \to \infty} \varepsilon = \frac{1}{2} \tag{2.166}$$

and

$$\lim_{N_S \to 0} \varepsilon = \frac{1 - e^{-N}}{2} \tag{2.167}$$

The highest efficiency is thus reached with the smallest switching period $N_{S1} \propto t_1$ or for the highest circulating stream $M_S c_{pS} \gg k_1 A_1$. Then, the storage temperature is constant and lies exactly between the inlet temperatures of streams 1 and 2. Therefore, efficiencies above 0.5 cannot be obtained (under these conditions) with the "short" regenerator, i.e., with $\vartheta_S \neq \vartheta_S(z)$. In order to approach the linear temperature profile of an ideal counterflow with equal capacities, a corresponding longitudinal profile $\vartheta_S(z)$ has to be set up in the storage mass. Longitudinal conduction in the solid ought to be kept as low as possible. This can be fulfilled to a high degree by increasing the thermal resistance in longitudinal conduction path, e.g., by gaseous gaps between the solid particles of a "classic" fixed bed regenerator or by thin walled material for the solid matrix of a rotating disc.

7.3 The "Long" Regenerator

The difference in the mathematical description of the long, compared with the short, regenerator is simply to be seen in the fact that the total first order derivative of the solid temperature with respect to time (or transverse coordinate) in eq. (2.158) has to be replaced by a partial derivative (just as in proceeding from crossflow over one row of tubes to ideal crossflow). Now, the solid temperature ϑ_S also depends on both independent variables [t (or y) and z] ($i = 1, 2$):

$$\boxed{-\frac{\partial \vartheta_S}{\partial \tau_i} = \vartheta_S - \vartheta_i} \tag{2.168}$$

$$\boxed{\frac{\partial \vartheta_i}{\partial \zeta_i} = \vartheta_S - \vartheta_i} \tag{2.169}$$

For the initial start-up of a regenerator, i.e., when the solid mass has a uniform temperature at $t = 0$ (or at $y = 0$, Fig. 2.46), the solution is exactly the same as for ideal crossflow (see eq. [2.51] and [2.52]). The crossflow profile established in the storage mass after the first (hot blow) period (or at the outlet edge $y = L_y$ of the first apparatus) now becomes the initial condition for the second (cold blow) period (or the entrance condition at the inlet edge of the second apparatus) and so forth. After sufficiently long operation, a periodic pattern of temperature profiles is established in the solid corresponding to the steady state of a rotating regenerator. The calculation of these temperature profiles is not possible in a simple closed form. The various mathematical methods for the solution of this problem are discussed in detail by Hausen [H2]. In the limiting case of short periods $N_{S1}, N_{S2} \to 0$, however, the efficiency of a long regenerator (see Fig. 2.46) can be given as that of a countercurrent cascade of short regenerators. In the case of equal capacities $R_1 = R_2$, for such a cascade of n equal short regenerators with efficiences ϵ_{sR}:

$$\frac{\epsilon}{1 - \epsilon} = n \frac{\epsilon_{sR}}{1 - \epsilon_{sR}} \tag{2.170}$$

At short switching periods ϵ_{sR} is to be calculated from eq. (2.167), using the NTU of the element, i.e., $(1/n)$th of the total NTU of the cascade:

$$\frac{\varepsilon}{1 - \varepsilon} = n \frac{1 - e^{-N/n}}{1 + e^{-N/n}} \tag{2.171}$$

Considering the limit of the cascade for $n \to \infty$ to be a continuous counterflow arrangement, one obtains (through series expansion of the exponential function)

$$\lim_{\substack{N_{S1} \to 0 \\ n \to \infty}} \frac{\varepsilon}{1 - \varepsilon} = \frac{N}{2} \tag{2.172}$$

The maximum efficiency of a "long" regenerator for short periods ($N_{S1,2} \to 0$) without longitudinal conduction thus becomes

$$\varepsilon_{\max} = \frac{N/2}{1 + N/2} \tag{2.173}$$

Let us further consider the fact that the overall heat transfer coefficient in N is defined for the transport only from the gas to the solid, while it is to be calculated from gas 1 to the wall and from the wall to gas 2 for a simple counterflow exchanger. Then, one can identify $(N/2)$ for the regenerator with the whole number of transfer units N of a corresponding gas-to-gas heat exchanger without intermediate storage. Even then the double surface area is still required, since, in N_1 and N, the area is always that of one of the two exchangers required in a regenerator. Figure 2.49 shows the efficiency of the regenerator versus N with the dimensionless period N_S as a parameter. The broken lines are calculated from eq. (2.165) for the "short" regenerator, i.e., with a solid temperature independent of the longitudinal coordinate (or an infinitely large heat

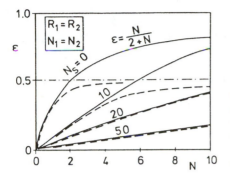

Figure 2.49 Efficiency ϵ of the regenerator vs. the NTU of a fluid stream N ($= N_1 = N_2$) with the nondimensional period N_S ($=$ NTU of the storage mass) as a parameter:—"long" regenerator (without) ---- "short" regenerator (with longitudinal conduction).

conductivity of the solid in z-direction). The full lines are obtained from numerical calculations [V1], while the uppermost curve for $N_S \rightarrow 0$ corresponds to eq. (2.173).

Since the most favorable case is obviously equivalent to the simple ideal counterflow (with $N/2 := N$), it is convenient to use an LMTD correction factor F here too. This enables one to design a regenerator exactly like a simple counterflow exchanger, where only the LMTD has to be corrected subsequently for $N_S > 0$. With equal capacities and equal exchangers, one can define

$$F = \frac{2}{N} \frac{\epsilon}{1 - \epsilon} \tag{2.174}$$

This F is identical to the quantity denoted as k/k_0 by Hausen [H2]. Hausen denotes N as a reduced length of the regenerator Λ and N_S as a reduced period Π. Hence, his diagram for k/k_0 (Λ, Π) that is also found in the VDI–WA [V1] and in the HEDH [H3] can also be applied to determine the efficiency. Numerical methods to calculate temperature profiles and efficiencies may be found in the same sources [H2, H3, V1].

Figure 2.50 shows a periodic pattern of solid temperatures for $N = 9$, $N_S = 10$ calculated using a finite difference scheme recommended by Hausen [H2]. Additionally, Hausen has given an approximation formula to calculate the LMTD correction

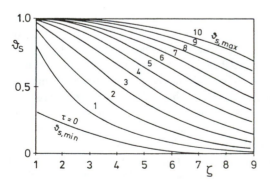

Figure 2.50 Periodic temperature pattern of storage mass (example).

factor, which reads

$$F = 1 - \frac{(4N_S/5) - 3\tanh(N_S/5)}{N} \qquad (2.175)$$

in the notation used here and is valid for $N_S/N = R < 0.5$. From this, we obtain

$$N_S = 1 : \qquad F_{(1)} = 1 - \frac{0.208}{N}$$

$$N_S = 2 : \qquad F_{(2)} = 1 - \frac{0.460}{N}$$

$$N_S = 3 : \qquad F_{(3)} = 1 - \frac{1.715}{N}$$

For switching periods used in practice, one can easily estimate now the deviation from ideal behavior for $N_S \to 0$ (see example chapter 3, section 4).

7.4 Thermal Coupling of Two Streams by Heat Pipes

A regenerator effect with $N_S \to 0$ can also be achieved by using heat pipes [D2]. The fluid in the heat pipe—which evaporates at the hot end, condenses at the cold end, and is recirculated to the hot end by gravity and/or capillary forces in a porous matrix (wick)—serves as the circulating heat carrier. Two gas streams in parallel ducts, for example, can be coupled very effectively by bundles of finned heat pipes [P1] and can be treated as a cascade of short regenerators with $N_S \to 0$.

8 CONCLUSIONS

8.1 Summary and Compact Presentation of the Formulas

The investigation of the influence of flow configuration on heat exchanger performance has shown, through the comparison of simple configurations, like stirred tank, parallel flow and counterflow, and the various crossflow configurations, that the type of flow configuration is crucial in the range close to thermal equilibrium ($N \to \infty$) (see Fig. 2.8). This is particularly so for equal or nearly equal flow capacities on both sides. For small number of transfer units ($N < 1$) or short relative residence times of the fluids in the apparatus (should NTU not be better interpreted as "nondimensional time unit"?), the transfer performance is affected far less by the flow configuration than by N itself.

Often the flow configurations occurring in real heat exchangers can be represented by cascades of interconnected "cells" or sub-exchangers, in each of which a simple configuration is realized. In such cases, the formulas that have been derived for the normalized mean temperature difference Θ or the normalized change in temperature ϵ of the simple configurations can be applied with good approximation, as shown in sections 3 to 7. It seems to be convenient, for a quick analysis of such equipment, to keep these formulas ready for application in a simple and compact form. For the design problem, the form $\Theta(\epsilon_1, \epsilon_2)$ would be best suited. As shown in Table 2.1, this function can be represented generally by the logarithmic mean of the temperature differences at both ends (subscripts "0", and "1") of the apparatus for the simplest configurations. The normalized temperature driving forces $\Delta\vartheta_0$ and $\Delta\vartheta_1$ and their differences are given in terms of the efficiencies ϵ_1, ϵ_2 below each sketch of the temperature profiles. If $\Delta\vartheta_0 = \Delta\vartheta_1$, as for counterflow with equal heat capacities and for stirred tank, both sides, then $\Theta = \Delta\vartheta_0 = \Delta\vartheta_1$. In these cases, the formula for Θ gives an indeterminate expression and leads to $\Theta = \Delta\vartheta_0 = \Delta\vartheta_1$ by series expansion.

The crossflow formulas in general cannot be solved explicitly for $\Theta(\epsilon_1, \epsilon_2)$. In these cases, one has to content oneself with a parametric representation $\Theta = (N_1, N_2)$, $\epsilon_i = N_i\Theta\epsilon_i = (N_1, N_2)$. Only for crossflow, one side mixed can the function $\Theta(\epsilon_1, \epsilon_2)$ be calculated explicitly (see eq. [2.29]).

The formulas for simple flow configurations are conventionally given in many text books in the $\epsilon(N, R)$ form. For asymmetric cases, such as crossflow, one side laterally mixed, two formulas are needed, depending on whether the reference NTU for the mixed stream is meant to be $N(= N_1)$ or $RN(= N_2)$. Writing the formulas in the form $\Theta(N_1, N_2)$ avoids the need for a priori definition of the reference NTU. Table 2.2 is a compilation of the formulas for the most important simple flow configurations. By multiplying Θ with N_1 or N_2, the required non-dimensional change in temperature ϵ_1 or ϵ_2 of the corresponding stream is easily obtained. In the asymmetric cases (stirred tank, one side, and crossflow, one side laterally mixed) the stream designated as 1 is regarded to be the mixed stream. If, for any reason, it is desired to address the mixed stream as 2, one has just to interchange the subscripts 1 and 2 in these formulas (see the results in section 1.1.2 where 2 was the mixed stream). All the others are symmetric cases, where interchanging the subscripts has no effect.

Table 2.1

Flow configuration	Stirred vessel (two-sided)	Stirred vessel (one-sided)	Parallel flow	Counter flow
Temperture profile, normalized: $\vartheta_1' = 1$ $\vartheta_2' = 0$				
$\Delta\vartheta_0$	$1-(\epsilon_1+\epsilon_2)$	$1-\epsilon_1$	1	$1-\epsilon_2$
$\Delta\vartheta_1$	$1-(\epsilon_1+\epsilon_2)$	$1-(\epsilon_1+\epsilon_2)$	$1-(\epsilon_1+\epsilon_2)$	$1-\epsilon_1$
$\Delta\vartheta_0 - \Delta\vartheta_1$	0	ϵ_2	$\epsilon_1+\epsilon_2$	$\epsilon_1-\epsilon_2$

Mean, normalized temperature difference for all four cases:

$$\Theta = \frac{\Delta\vartheta_0 - \Delta\vartheta_1}{\ln\dfrac{\Delta\vartheta_0}{\Delta\vartheta_1}} \quad (\Delta\vartheta_0 \neq \Delta\vartheta_1)\ ;\quad \Theta = \Delta\vartheta_{0,1} \quad (\Delta\vartheta_0 = \Delta\vartheta_1)$$

the mixed stream). All the others are symmetric cases, where interchanging the subscripts has no effect.

In Table 2.2, one often comes across a term such as $N/(1 - e^{-N})$. In order to write the formulas in a more compact form, this term may be denoted as a function

$$\varphi(X) \equiv \frac{X}{1 - e^{-X}} \tag{2.176}$$

As may be seen from the right hand column of Table 2.2, substituting N_1, N_2, the sum $(N_1 + N_2)$ or the difference $(N_1 - N_2)$, respectively, for the variable X, extremely compact expressions can be obtained for the reciprocal of the normalized mean tem-

Table 2.2

Configuration	$\Theta(N_1, N_2)$ $\varepsilon_i = N_i\Theta$	$\dfrac{1}{\Theta(N_1, N_2)}$ *
Stirred vessel, both sides	$\dfrac{1}{1 + N_1 + N_2}$	$1 + N_1 + N_2$
Stirred vessel, one side (1)	$\dfrac{1}{N_1 + N_2/(1 - e^{-N_2})}$	$N_1 + \varphi(N_2)$
Parallel flow	$\dfrac{1 - e^{-(N_1 + N_2)}}{N_1 + N_2}$	$\varphi(N_1 + N_2)$
Counterflow	$\dfrac{1 - e^{-(N_1 - N_2)}}{N_1 - N_2 e^{-(N_1 - N_2)}}$ $(N_1 \neq N_2)$ $\dfrac{1}{1 + N}$ $(N_1 = N_2)$	$\varphi(N_1 - N_2) + N_2$
Crossflow, both sides laterally mixed	$\left[\dfrac{N_1}{1 - e^{-N_1}} + \dfrac{N_2}{1 - e^{-N_2}} - 1 \right]^{-1}$	$\varphi(N_1) + \varphi(N_2) - 1$
One side (1) laterally mixed	$\dfrac{1 - \exp[-N_1(1 - e^{-N_2})/N_2]}{N_1}$	$\varphi\left(\dfrac{N_1}{\varphi(N_2)} \right) \varphi(N_2)$
Unmixed	$\Theta = \displaystyle\sum_{m=0}^{\infty} \dfrac{1 - \sum_{n=0}^{m} e^{-N_1} N_1^n/n!}{N_1} \cdot \dfrac{1 - \sum_{n=0}^{m} e^{-N_2} N_2^n/n!}{N_2}$	

* Short form using the auxiliary function $\varphi(X) = X/(1 - e^{-X})$

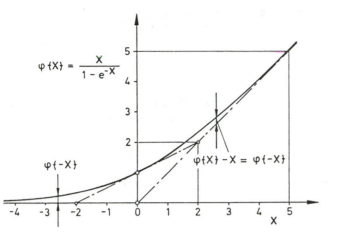

Figure 2.51 Graph of auxiliary function $\varphi(X)$.

perature difference Θ. This does not work for unmixed (ideal) crossflow, but even there one can recognize that the terms of the series are a form similar to $1/\varphi(N_i)$. It may be recalled that the algorithm for the evaluation of this series has been given earlier in Fig. 2.7. The characteristics of the function $\varphi(X)$ are shown in Fig. 2.51.

At $X = 0$, $\varphi(X)$ has a limiting value of unity,

$$\boxed{\varphi(0) = 1} \tag{2.177}$$

which may be seen from a series expansion of the exponential function. Its slope at $X = 0$ is 1/2, and it tends to its argument for large values of X

$$\boxed{\varphi(X) \to X \qquad (X \gg 1)} \tag{2.178}$$

Thus, the compact formulas in Table 2.2 and their limiting cases are very easily handled in practice. For example, one can find the normalized temperature change ϵ_i of both streams in *parallel flow* from the very simple formulas

$$\varepsilon_1 = \frac{N_1}{\varphi(N_1 + N_2)} \tag{2.179}$$

and

$$\varepsilon_2 = \frac{N_2}{\varphi(N_1 + N_2)} \tag{2.180}$$

and their limiting values may be immediately arrived at, if the behavior of $\varphi(X)$ is kept in mind. Here, the argument X is the sum of the NTUs, and the physical meaning of φ is the ratio of the maximum temperature difference $(\vartheta_1' - \vartheta_2')$ to the mean temperature difference $(\vartheta_1 - \vartheta_2)_M$ for parallel flow. For *counterflow*, the difference of the NTUs replaces the sum, but the expression could be written as a vector sum

$$\frac{1}{\Theta} = \varphi(\vec{N}_1 + \vec{N}_2) - \vec{N}_2$$

Note that in Table 2.2 the N_i are meant to be absolute values. While two formulas, for $N_1 \neq N_2$ and $N_1 = N_2$, are needed in the conventional way of writing, as in the left-hand column, a single and much simpler one suffices in the new way (right-hand column). The efficiencies are now obtained from

$$\varepsilon_1 = \frac{N_1}{\varphi(N_1 - N_2) + N_2} \tag{2.181}$$

and

$$\varepsilon_2 = \frac{N_2}{\varphi(N_2 - N_1) + N_1} \tag{2.182}$$

The counterflow limiting cases are $C = -1$ and $C = 0$. For $C = -1$, the flow capacities are equal, $N_1 - N_2 = 0$, and ϵ for this case, with $\varphi(0) = 1$, is

$$\varepsilon = \frac{N}{1 + N} \qquad (C = -1) \tag{2.183}$$

For $C = 0$, N_2 is also zero since $N_2 = |C|N_1$, and we get

$$\varepsilon_1 = \frac{N_1}{\varphi(N_1)} \qquad (C = 0) \tag{2.184}$$

As the flow direction is of no importance when $C = 0$, the same result can be found for parallel flow from eq. (2.179). Note that, due to the property of the function $\varphi(X)$ according to

$$\varphi(X) = \varphi(-X) + X \tag{2.185}$$

the subscripts in the Θ-formula for counterflow can be freely interchanged. Therefore, the denominators in eq. (2.181) and (2.182) are always equal

$$\varphi(N_1 - N_2) + N_2 = \varphi(N_2 - N_1) + N_1 \tag{2.186}$$

The formula for *crossflow, one side laterally mixed,* results in the efficiency expression

$$\varepsilon_1 = \frac{N_1/\varphi(N_2)}{\varphi(N_1/\varphi(N_2))} \tag{2.187}$$

For equal flow capacities, $N_1/\varphi(N_2)$ can reach a maximum value of unity for $N_1 = N_2 \to \infty$, and its maximum efficiency becomes

$$\varepsilon_{max} = \frac{1}{\varphi(1)} = 0.632 \qquad (2.188)$$

The function φ also figures in the case of *crossflow over n rows of tubes*. The short notations g (eq. [2.36]) and $b = gN_2/n$ that were introduced in section 2.2 for this case can be seen to be related to φ. Writing further abbreviated symbols a, b, B, and Z as

$$a = e^{-N_2/n} \qquad (2.189)$$

$$b = 1 - a = \frac{N_2/n}{\varphi(N_2/n)} \qquad (2.190)$$

$$B = \frac{N_1}{\varphi(N_2/n)} = \frac{nb}{R} \qquad (2.191)$$

$$Z = bB = \frac{nb^2}{R} \qquad (2.192)$$

we can put down the efficiency formulas for number of tube rows $n = 1$ to 6 in tabular form:

n	$(1 - \varepsilon_1)e^B$	crossflow over n row of tubes
1	1	
2	$1 + \dfrac{1}{2}Z$	(2.193)
3	$1 + \dfrac{1}{3}[(a+2)Z + Z^2/2!]$	
4	$1 + \dfrac{1}{4}[(a^2 + 2a + 3)Z + (2a+2)Z^2/2! + Z^3/3!]$	
5	$1 + \dfrac{1}{5}[(a^3 + 2a^2 + 3a + 4)Z + (3a^2 + 4a + 3)Z^2/2! + (3a+2)Z^3/3! + Z^4/4!]$	
6	$1 + \dfrac{1}{6}[(a^4 + 2a^3 + 3a^2 + 4a + 5)Z + (4a^3 + 6a^2 + 6a + 4)Z^2/2! +$ $+(6a^2 + 6a + 3)Z^3/3! + (4a+2)Z^4/4! + Z^5/5!]$	

A solution for an arbitrary number of tube rows is given in Appendix A. The properties of the function $\varphi(X)$ reveal that, as $N_2 \to 0$, b and Z tend to zero while

$B \to N_1$. Thus, all the terms except unity in the set of eqs. (2.193) vanish, and the correct result $\epsilon_1(N_2 \to 0) = 1 - e^{-N_1}$ is obtained. The validity of the result is obvious since the heat capacity rate goes to zero for $N_2 \to 0$ and the temperature of stream 2 remains constant. More interesting is the result corresponding to the other limit $N_2 \to \infty$ at fixed flow capacity ratio R. As may be seen again from the behavior of φ, in this limit the quantities a, b, Z, and B tend to the values

$$
\lim_{N_2 \to \infty}
\begin{pmatrix}
a \\
b \\
Z \\
B
\end{pmatrix}
=
\begin{pmatrix}
a_\infty = 0 \\
b_\infty = 1 \\
Z_\infty = n/R \\
B_\infty = n/R
\end{pmatrix}
\tag{2.194}
$$

So, all terms containing a vanish in this limit and finding the asymptotic values of the efficiencies $\epsilon_{\infty,n}$, is easier. At equal capacities on both sides ($R = 1$), these values for $n = 1$ to 6 yield

$$\varepsilon_{\infty,1} = 1 - e^{-1} \qquad = 0.632$$

$$\varepsilon_{\infty,2} = 1 - 2e^{-2} \qquad = 0.729$$

$$\varepsilon_{\infty,3} = 1 - \frac{9}{2} e^{-3} \qquad = 0.776$$

$$\varepsilon_{\infty,4} = 1 - \frac{64}{6} e^{-4} \qquad = 0.805$$

$$\varepsilon_{\infty,5} = 1 - \frac{625}{24} e^{-5} \qquad = 0.825$$

$$\varepsilon_{\infty,6} = 1 - \frac{7776}{120} e^{-6} = 0.839$$

Using the relation $n^{n-1}/(n - 1)! = n^n/n!$, these first six values calculated may be generalized as

$$\varepsilon_{\infty,n} = 1 - \frac{n^n}{n!} e^{-n} \tag{2.195}$$

This result can also be derived in a general way for arbitrary number of rows n from eq. (2.34) and (2.40) as shown in Appendix A. For larger n, the calculation can be considerably simplified using Stirling's formula for the factorials of large numbers:

$$\varepsilon_{\infty,n} \approx 1 - \frac{1}{\sqrt{2\pi n}} \tag{2.196}$$

Against the number of tube rows n, Fig. 2.52 plots the limiting efficiency $\epsilon_{\infty,n}$ calculated from eq. (2.195), as well as from eq. (2.196), using Stirling's formula for $n!$. In practice, the asymptotic approximation of eq. (2.196) is sufficient for $n > 4$. The

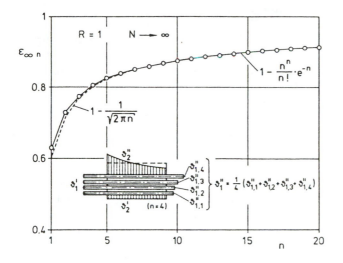

Figure 2.52 Maximum possible efficiencies $\epsilon_{\infty,n}$ for cross flow with n rows of tubes at equal flow capacities ($R = 1$, $N \to \infty$).

maximum possible value of unity, to be reached for ideal crossflow, requires an infinite heat transfer area in infinite number of tube rows. From Fig. 2.52, the approach to this efficiency $\epsilon_{\infty,\infty}$ is seen to be very slow. From eq. (2.196), we can calculate the number of tube rows in crossflow at equal capacity rates required to achieve a given efficiency:

$$n(\varepsilon_{\infty,n}) = \frac{1}{2\pi(1 - \varepsilon_{\infty,n})^2} \tag{2.197}$$

To obtain ϵ_∞ of 90%, 95%, and 99%, the corresponding number of tube rows are determined to be 16, 64, and 1600! From this, one can see the practical implications of designing crossflow heat exchangers having high efficiency, as, for example, in heat recovery. In such cases, the answer lies in a multiple cross-counterflow arrangement.

For an apparatus with *two passes* and many baffles, typified by a shell-side stream laterally mixed, the compact $1/\Theta$ form can be written using the function φ:

$$\boxed{\frac{1}{\Theta} = \varphi(N_{12}) + \frac{1}{2}(N_1 + N_2 - N_{12})} \tag{2.198}$$

$$\boxed{N_{12} = (N_1^2 + N_2^2)^{1/2}}$$

With the notation used earlier for this, we can now write eq. (2.88) as

$$\boxed{\varepsilon = \frac{2}{[2\varphi(\omega N)/N] + (1 + R - \omega)}} \tag{2.199}$$

$$\omega = (1 + R^2)^{1/2}$$

Here, too, the limiting value ϵ_∞ of eq. (2.89) for $N \to \infty$ is derived, considering that $\varphi(\omega N)/N \to \omega$ in the limit. The passes being connected one in parallel flow and the other in counterflow, the argument of the function turns out to be the root mean square of the sum (for parallel flow) and difference (for counterflow) of the NTUs:

$$N_{12} = [N_1^2 + N_2^2]^{1/2} = \left\{ \frac{1}{2}[(N_1 + N_2)^2 + (N_1 - N_2)^2] \right\}^{1/2} \qquad (2.200)$$

For numbers of tube passes greater than two in a laterally mixed shell-side stream, the formulas become more complex. As shown by Gardner [G5, G6] and Hausen [H2], the limiting case of an infinite number of tubeside passes in a laterally mixed shell-side stream is treated very easily, it being identical with crossflow both sides laterally mixed (why indeed?)!

At low values of NTU, the efficiencies for two passes and an infinite number of passes are not very different. At $R = 1$,

$$\epsilon_{2 \text{ passes}} = \frac{N}{\varphi(\sqrt{2}N) + (1 - \sqrt{2}/2)N} \qquad (2.201)$$

and

$$\epsilon_{\infty \text{ passes}} = \frac{N}{2\varphi(N) - 1} \qquad (2.202)$$

With $\varphi(N) = 1 + (N/2)$ for $N \ll 1$, both equations give $\epsilon = N/(1 + N)$. For an infinite number of passes ($=$ crossflow, both sides mixed), the efficiency ϵ traverses a maximum value of $\epsilon_{\text{max}, \infty \text{ passes}} = 0.5645$ (at $N \approx 3$, for $R = 1$) and decreases to $\epsilon_{\infty, \infty \text{ passes}} = 0.5$ for $N \to \infty$. Figure 2.53 shows the efficiencies of heat exchangers with 2, 3, 4, and an infinite number of tubeside passes and one shell-side pass, laterally mixed, at equal flow capacities plotted versus the mean temperature difference Θ with N as a parameter (see upper right quarter of Fig. 2.8 for comparison).

From this, it can be found that the multipass heat exchangers with even numbers of internal passes lie in the lower crossflow region between curve d for crossflow, both sides mixed, and the curve for $n = 2$ from eq. (2.201). For odd numbers of internal passes, the efficiency is higher or lower than for the next lower even number, depending on flow direction (more counterflow passes are, of course, better). The analysis for *three passes* is somewhat cumbersome, as one has to solve four coupled differential equations and determine 16 constants from these equations and the boundary conditions. Fischer [F1] has given an analytical solution for the case where two of the three passes are in counterflow with respect to the shell-side stream. Fischer's solution, written in the new way $1/\Theta$ (N_X, N_Y) reads

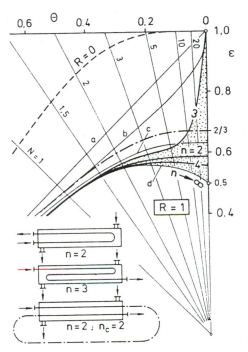

Figure 2.53 Efficiency ϵ and normalized mean temperature difference Θ for multiphass heat exchangers with laterally mixed shell-side stream at equal flow capacities $R = 1$ —.—.—. ($n = 2$, $n_c = 2$).

$$\frac{1}{\Theta} = \frac{N_X - N_Y}{f(N_X, N_Y) - 1} + N_X \qquad (n = 3, R \neq 1) \qquad (2.203)$$

$$f(N_X, N_Y) = \frac{[2 - a/\varphi(a)]\varphi(Z) - a - 4c}{[2 - b/\varphi(b)]\varphi(Z) - b + 4c} \quad \frac{\varphi(-c)}{\varphi(+c)}$$

$$Z = \left(N_X^2 + \frac{4}{9}N_Y(N_Y - N_X) \right)^{1/2}$$

$$a = \frac{Z}{2} - \frac{N_X}{2} - \frac{N_Y}{3} \qquad b = Z - a \qquad c = \frac{N_Y}{3}$$

The subscript X here denotes the outer, laterally mixed stream, which is in counterflow to two of the three inner passes (stream Y) (see Fig. 2.53). Due to the asymmetry here, we did not use the subscripts 1 and 2. For equal flow capacities, $R = 1$, i.e., $N_Y = N_X$, eq. (2.203), as for pure counterflow, leads to an indeterminate expression. The limit is not obvious from this somewhat more complicated function.

In Appendix B, a rigorous solution of the problem for the specific case $R = 1$ is derived from the fundamental equations. With this, too, the correctness of Fischer's calculations can be checked. Now in place of eq. (2.203), one obtains

$$\frac{1}{\Theta} = N + \frac{9N}{N + 8f_1(N)} \qquad (n = 3, R = 1)$$

$$f_1(N) = \frac{1 + x - x^3 - x^4}{1 + x^4} \qquad x = e^{-N/3} \qquad (2.204)$$

The values of the LMTD correction F tabulated by Fischer [F1] as a function of ϵ, deviate from those calculated from eq. (2.204) ($F = \Theta/\Theta_{LM}$ with $\Theta_{LM} = 1 - \epsilon$ at $C = -1$) by up to 7.5%. At $\epsilon = 0.603$, F is 0.416. According to Fischer, however, it is 0.450. The reason is that he could calculate the limit of equal capacities only approximately from his formulas with $R = 0.999$ or $R = 1.001$, for example. The asymmetry of eq. (2.203) shows itself if the normalized mean temperature difference and the efficiency are calculated first for $N_Y = 15$, $N_X = 10$ ($R = 2/3$) and then for $N_Y = 10$, $N_X = 15$ ($R = 3/2$). In the first case (stronger shell-side stream), one obtains $\Theta = 0.0570$ and $\epsilon_Y = 0.856$ ($\epsilon_X = 0.570$) and, in the second case, $\Theta = 0.0520$ and $\epsilon_X = 0.780$ ($\epsilon_Y = 0.520$).

The curve shown in Fig. 2.53 for $n = 3$ was calculated from eq. (2.204). While the differences to the cases of two or an infinite number of passes remain comparatively small for NTU below about four, they become significant at very high NTU. In the limit, the dominant counterflow behavior comes to light; and, in principle, one can reach maximum efficiencies of $\epsilon_\infty \to 1$ with such a configuration. The results of Fischer's calculations were shown only up to $F = 0.450$, i.e., up to a maximum N of about 3.4 at $R = 1$, so that this interesting behavior at large NTU could not be recognized. The LMTD correction does not tend to zero as it does for two passes, but it reaches an asymptotic value of $F_\infty = 1/9$, as can be derived from eq. (2.204). Here ($n = 3$, $C = -1$) the same efficiency can only be reached in the limit with nine times the number of transfer units required in an ideal counterflow exchanger. But contrary to two passes, efficiencies above 0.6 are, in principle, not impossible at equal capacities.

It seems to be logical to increase the efficiency by insulating the middle parallel flow pass, i.e., to transform it into a pure bypass with no heat transfer and, thus, come to an apparatus with *two counterflow passes*. Recently, Roetzel [R3] has presented calculations and design ideas for such an apparatus. The corresponding formula for an apparatus with two internal counterflow passes and one shell-side stream can be written rather simply in the new way as (problem!)

$$\frac{1}{\Theta} = \varphi\left(N_X - \frac{N_Y}{2}\right) + \frac{N_Y}{2}\left(1 + \frac{\varphi(N_Y)}{2\varphi(N_Y/2)}\right) \qquad (n = 2, n_c = 2) \quad (2.205)$$

$$\varepsilon_{X,Y} = N_{X,Y}\Theta \qquad (2.206)$$

At $R = N_Y/N_X = 2$, the argument of the first term becomes zero and $\varphi(0) = 1$. The expression in the second term varies between 3/2 and 2 for $N_Y \to 0$ and $N_Y \to \infty$, respectively. At the particular value of $R = 2$, the apparatus has the same asymptotic efficiency ϵ_Y as an ideal counterflow exchanger at $R = 1$ [$\epsilon = N/(1 + N)$].

As an energy balance will show, the corresponding efficiency ϵ_X of the stronger stream cannot surpass the value of 0.5. However, the apparatus with two counterflow passes at $R = 1$ is not at all an ideal counterflow exchanger! Its efficiency is indeed higher than that of a crossflow exchanger, one side laterally mixed, but it reaches a terminal value far below unity: ϵ_∞ at $R = 1$ is 2/3 as shown by the dotted curve in Fig. 2.53. At very high N and equal heat capacities, it would, therefore, make little sense to insulate the parallel flow pass in an apparatus with $n = 3$.

As in the case of $n = 2$, for *four passes* an analytical solution has also been found by Underwood [U1, F1] in the 1930s. In the $1/\Theta$ form, it can be written in a much simpler way than for three passes (problem!).

With $X = N_X$, $Y = N_Y$, and $Z = [X^2 + (Y/2)^2]^{1/2}$, one obtains

$$\frac{1}{\Theta} = \varphi(Z) + \varphi(Y) - \varphi\left(\frac{Y}{2}\right) + \frac{1}{2}\left(X + \frac{Y}{2} - Z\right) \quad (n = 4) \qquad (2.207)$$

For infinite capacity flow rate and, hence, no temperature change of the X-side fluid, X becomes zero, and Θ should be the well-known function of only Y; likewise, for $Y = 0$. It is easily checked that eq. (2.207) satisfies these conditions: $1/\Theta = \varphi(Y)$ for $X = 0$ and $1/\Theta = \varphi(X)$ for $Y = 0$. In its original form as given by Underwood, the mean temperature difference was represented in terms of hyperbolic tangents and cotangents. These are related to the auxiliary function $\varphi(x)$ by:

$$\tanh x = \frac{\varphi(4x)}{x} - \frac{\varphi(2x)}{x} - 1$$

$$\coth x = \frac{\varphi(2x)}{x} - 1$$

Actually, $n = 4$ is a weakly asymmetric case: with $Y = 5$, $X = 10$, one finds $1/\Theta = 13.715$, $\epsilon_X = 0.7292$, and $Y = 10$, $X = 5$ leads to $1/\Theta = 13.507$, $\epsilon_Y = 0.7403$, a difference of 1.5%. In most cases, the asymmetry is even less. At $Y = X$ ($R = 1$), the asymptotic efficiency for $N \to \infty$ ($Y = X = N$, $Z = \sqrt{5}\,N/2$) becomes

$$\varepsilon_\infty(R = 1, n = 4) = \frac{4}{5 + \sqrt{5}} = 0.5528 \qquad (2.208)$$

about 6% below the corresponding value for $n = 2$. This is not the maximum, however, for $R = 1$; at $N \approx 3.25$, one finds $\epsilon_{max} = 0.5691$ (see Fig. 2.53).

For the *series-parallel arrangements* that are frequently encountered in plate heat exchangers (see section 4.3), the efficiencies are suitably calculated from a generalized form of eq. (2.121):

$$(1 - \varepsilon_Y)_{1 \times n} = (1 - \varepsilon_{p\,Y})^{n_p}(1 - \epsilon_{c\,Y})^{n_c} \qquad (2.209)$$

$$\varepsilon_{p\,Y} = \frac{Y/n}{\varphi((Y/n) + X)} \qquad \text{(Parallel flow)}$$

$$\varepsilon_{c\,Y} = \frac{Y/n}{\varphi((Y/n)-X)+X} \qquad \text{(Counterflow)}$$

Stream Y flows through the apparatus in n-fold series connection, while stream X is subdivided into n parallel substreams (see Fig. 2.54). The exponents n_p and n_c are the number of parallel and counterflow passes, respectively. For even numbers n, we have $n_p = n_c = n/2$. For odd numbers n, the more favorable $n_p = (n-1)/2$ and $n_c = (n+1)/2$ should be chosen. If the capacity rate ratio is equal to the number of passes, $R = n$, and $Y = nX$ in the formula for ε_{cY}. we again have $\varphi(0) = 1$ and two different formulas, as required in eq. (2.117), are not needed. The maximum possible efficiencies of such series-parallel arrangements are easy to recognize from this as

$$\varepsilon_{p\,Y,\,max} = \frac{1}{1+n/R} \tag{2.210}$$

$$\varepsilon_{c\,Y,\,max}(R \ge n) = 1, \qquad \varepsilon_{c\,Y,\,max}(R < n) = \frac{R}{n} \tag{2.211}$$

Thereby, the maximum efficiency of the whole arrangement is always

$$\varepsilon_{Y,1\times n,\,max}(R \ge n) = 1$$

$$\varepsilon_{Y,1\times n,\,max}(R < n) = 1 - \left(1-\frac{R}{n}\right)^{n_c}\left(1-\frac{1}{1+n/R}\right)^{n_p} \tag{2.212}$$

For example, with $n = 2$, i.e., $n_c = n_p = 1$, and $R = 1$, the maximum efficiency is

$$\varepsilon_{1\times2,\,max}(R = 1) = 1 - \frac{1}{2}\frac{2}{3} = \frac{2}{3} = 0.667 \tag{2.213}$$

and for $n = 3$ with $n_c = 2$, $n_p = 1$ and $R = 1$ also, it is the same value:

$$\varepsilon_{1\times3,\,max}(R = 1, n_c = 2) = 1 - \left(\frac{2}{3}\right)^2\frac{3}{4} = \frac{2}{3} = 0.667 \tag{2.214}$$

For the less favorable case $n_c = 1$, $n_p = 2$, it follows that

$$\varepsilon_{1\times3,\,max}(R = 1, n_c = 1) = 1 - \frac{2}{3}\cdot\left(\frac{3}{4}\right)^2 = \frac{5}{8} = 0.625 \tag{2.215}$$

As n increases, the maximum efficiency at $R = 1$ begins to decrease gently, and, in the limit $n \to \infty$, the lowest value of ε_{max} at $R = 1$ is reached:

$$\lim_{n \to \infty}\left(1-\frac{1}{n}\right)^n = e^{-1}$$

$$\varepsilon_{1\times\infty,\,max}(R = 1) = 1 - e^{-1} = 0.632 \tag{2.216}$$

This coincides with the corresponding value for crossflow, one side mixed, eq. (2.188).

The $1 \times n$ series-parallel arrangement terminates in crossflow, one side mixed (see Fig. 2.53 curve c) just as the arrangement of n inner passes in a laterally mixed outer stream leads to crossflow, both sides mixed (see Fig. 2.53 curve d) as a lower limit for $n \to \infty$. The two crossflow limits have already been discussed by Gardner [G5, G6] in 1941.

The development of these two fundamental families of flow configurations, from parallel or counterflow to crossflow, one side or both sides laterally mixed, is shown schematically in Fig. 2.54. The differences and similarities of various flow configurations may be best recognized from such simple symbolic figures. The multiple passes are denoted by the number of passes n, the number of parallel flow passes n_p, and the number of counterflow passes n_c. One of these three could have been omitted since $n_p + n_c$ is always equal to n. The corresponding calculation formulas are again compiled in Table 2.3 in a unified and compact manner. As before (eq. [2.207]), X here stands for the NTU of the outer stream and Y, the NTU of the inner stream. The NTU may indeed be regarded as the terminal values of a nondimensional length coordinate ($0 \le x \le X$). Table 2.3 contains expressions for $1/\Theta$ (X, Y), which can be formulated very compactly with the auxiliary function $\varphi(x)$, with the exception of the slightly more complicated case with three passes. The normalized changes in temperature or efficiencies ϵ are again easily obtained from $\epsilon_x = X\Theta$ or $\epsilon_Y = Y\Theta$. For the exchangers with even numbers of passes $n = 2m$ and $n_p = m$, $n_c = m$, one can even find a general equation which is in agreement with the results of the analyses for $n = 2$, $n = 4$, and $n \to \infty$ (see Table 2.3). From this the asymptotic efficiencies of the $(2m, m, m)$ arrangements are

$$\varepsilon_\infty(R = 1) = \frac{1}{1.5 + (\sqrt{1 + m^2} - 1)/(2m)} \tag{2.217}$$

$$\varepsilon_{Y, \infty}(R = 2m) = \frac{1}{1 + (\sqrt{5} - 1)/(2m)} \tag{2.218}$$

In fact, the generalized equation for $(2m, m, m)$ given in Table 2.3 is mathematically correct also for $2m = 6, 8, 10$, etc., passes, as may be seen from a comparison (see Appendix C) with a generalized solution obtained by Kraus and Kern in 1965 [K4]. (This reference has been brought to my attention only recently by B. Spang, from Wilfried Roetzel's laboratory in Hamburg).

Without rewriting the known solutions for two and four passes in the new way, a simple closed form of the generalized solution for $(2m, m, m)$ would probably not have been found.

Since lateral mixing of the shell-side streams lowers the efficiency, one can obtain a connection in parallel in baffled shell-and-tube heat exchangers by using a suitable longitudinal shell-side baffle [G6]. These configurations with multiple passes on the shell side may be treated by the equations for countercurrent cascades and series-parallel arrangements.

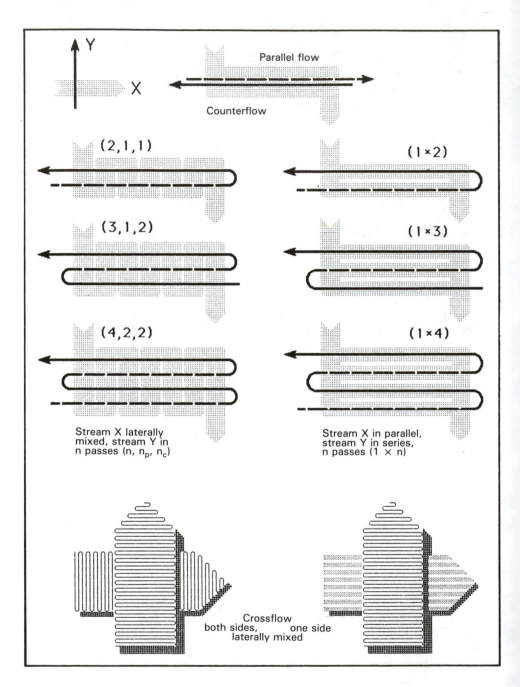

Figure 2.54 Heat exchangers with *n* passes in a laterally mixed shell-side stream and series-parallel arrangements (scheme, see Table 2.3).

Table 2.3

$$X = \frac{kA}{(\dot{M}c_p)_X}, \qquad Y = \frac{kA}{(\dot{M}c_p)_Y}, \qquad \varepsilon_X = X\Theta, \qquad \varepsilon_Y = Y\Theta, \qquad \varphi(X) = \frac{X}{1 - e^{-X}}$$

n n_p n_c	$\dfrac{1}{\Theta(X,Y)}$ $\rightarrow \rightarrow$ $\rightarrow \leftarrow$ for n internal passes in laterally mixed outer stream		ε_∞ $(Y = X)$	$\varepsilon_{Y,\infty}$ $(Y = nX)$
1 1 0	$\varphi(X + Y)$	Parallel flow	0.5	0.5
1 0 1	$\varphi(X - Y) + Y$	Counterflow	1.0	1.0
2 1 1	$\varphi(Z_1) + \dfrac{X + Y - Z_1}{2}$	$Z_1 = [X^2 + Y^2]^{1/2}$	0.586	0.764
4 2 2	$\varphi(Z_2) + \varphi(Y) - \varphi\left(\dfrac{Y}{2}\right) + \dfrac{X + Y/2 - Z_2}{2}$	$Z_2 = \left[X^2 + \left(\dfrac{Y}{2}\right)^2\right]^{1/2}$	0.553*	0.866
$2m$ m m	$\varphi(Z_m) + \varphi(Y) - \varphi\left(\dfrac{Y}{m}\right) + \dfrac{X + Y/m - Z_m}{2}$	$Z_m = \left[X^2 + \left(\dfrac{Y}{m}\right)^2\right]^{1/2}$		
∞ ∞ ∞	$\varphi(X) + \varphi(Y) - 1$	Crossflow, both sides laterally mixed	0.5**	1.0
2 0 2	$\varphi\left(X - \dfrac{Y}{2}\right) + \dfrac{Y}{2}\left[1 + \dfrac{1}{2}\dfrac{\varphi(Y)}{\varphi(Y/2)}\right]$		0.667	1.0
3 1 2	$X + \dfrac{X - Y}{f(X,Y) - 1}$	$Z = \left[X^2 + \dfrac{4}{9}Y(Y - X)\right]^{1/2}$	1.0	1.0
	$f(X,Y) = \dfrac{[2 - a/\varphi(a)]\,\varphi(Z) - a - 4c}{[2 - b/\varphi(b)]\,\varphi(Z) - b + 4c}\dfrac{\varphi(-c)}{\varphi(+c)}$ $a = \dfrac{Z}{2} - \dfrac{X}{2} - \dfrac{Y}{3} \qquad b = Z - a \qquad c = \dfrac{Y}{3}$			
n n_p n_c	$\dfrac{1}{\Theta(X,Y)}$ for series-parallel arrangements $(1 \times n)$			
n n_p n_c	$\dfrac{Y}{1 - (1 - \varepsilon_p)^{n_p}(1 - \varepsilon_c)^{n_c}}$			
	$\varepsilon_p = \dfrac{Y/n}{\varphi(Y/n + X)} \qquad \varepsilon_c = \dfrac{Y/n}{\varphi(Y/n - X) + X}$			
∞ ∞ ∞	$\varphi\left(\dfrac{Y}{\varphi(X)}\right)\varphi(X)$	Crossflow, one side*** laterally mixed	0.632	1.0

* $\varepsilon_{max}(X = Y = 3.25) = 0.569$ ** $\varepsilon_{max}(X = Y = 3) = 0.565$ *** (Y-side)

Equation 2.142 for the efficiency of a *pair of heat exchangers, coupled* by a circulating heat carrier ("S") (see Fig. 2.41) may be written in the form $1/\Theta$ (X, Y, Z_1, Z_2):

$$\frac{1}{\Theta} = \frac{1}{\Theta(X, Z_1)} + \frac{Z_1/Z_2}{\Theta(Y, Z_2)} - Z_1 \qquad (2.219)$$

$$X = \left(\frac{kA}{\dot{M}c_p}\right)_1, \quad Y = \left(\frac{kA}{\dot{M}c_p}\right)_2, \quad Z_1 = \frac{(kA)_1}{(\dot{M}c_p)_S}, \quad Z_2 = \frac{(kA)_2}{(\dot{M}c_p)_S}$$

This is also valid for the "short" regenerator (see eq. [2.146]). With equal exchangers ($Z_1 = Z_2 = Z$), this form is simpler than the one written with ϵ. The function $\Theta(X, Y)$—or its reciprocal—is generally suited for a compact representation of the relationships between the transfer performances and the residence times of the fluids in an apparatus. The function $\Theta(\epsilon_X, \epsilon_Y)$, which would be more suitable for the design problem, is not available in an explicit form in most cases. Roetzel and Nicole [R2] have, therefore, suggested an empirical approximation formula for $\Theta(\epsilon_X, \epsilon_Y)$ with 16 constants that can be fitted to the exact implicit calculation. The constants so determined have been given along with the diagrams $\Theta(\epsilon_X, \epsilon_Y)$ in the fifth edition of VDI–Waermeatlas. Perhaps it is simpler and safer to use the exact implicit form $\Theta(X, Y)$, $\epsilon(X, Y)$ from Table 2.3, as the copying of 16 constants is fraught with the risk of errors.

8.2 Premises and Limitations of Linear Theory

A common feature of all the analyses and, therefore, all the results of the first and second chapters was the simple linear law for the kinetics of heat transfer from the hot to the cold medium (eqs. [1.12, 1.49, 1.59, 1.60, 1.81, 1.82, 2.16, 2.17, 2.49, 2.50, 2.58, 2.59]) and the energy balance (e.g., eqs. [1.50, 1.51]), in which only the change in enthalpy in the flow direction and the heat flux perpendicular to it are accounted for in steady state. The validity of the results is, therefore, necessarily restricted to the range of validity of these fundamental equations. Moreover, it has been assumed, during integration of the fundamental equations, that the overall heat transfer coefficients k and the specific heat capacities c_p of the fluids are independent of temperature or temperature difference. This does not imply, however, as often stated in papers and books in this context, that the whole "linear theory" would be valid only for constant overall heat transfer coefficients k. The overall heat transfer coefficient k may indeed depend on the flow path z, as it actually does to a high degree in laminar flow, for example, while the length dependence is restricted to a short developing zone for turbulent flow. In the equations to calculate the outlet temperatures or the mean temperature difference, the overall heat transfer coefficient is, then, an integral average over the whole surface area A

$$k = \frac{1}{A} \int_0^A k_{loc}(A)\, dA \qquad (2.220)$$

From measurements on industrial heat exchangers, usually only these average values of k can be determined. If the distribution $k_{loc}(A)$ is not known, one certainly cannot calculate the exact variation of temperatures inside the apparatus. Nevertheless, the linear theory is still correct, within the framework of its premises, if the calculation is restricted to the determination of outlet temperatures. Usually the following additional processes are not accounted for in the balances: heat exchange with the surroundings ("heat losses"), heat conduction in flow direction in the fluid and in the separating walls ("longitudinal conduction"), dissipation of mechanical pumping power, and expansion or compression power of the fluids. The influence of these energy terms, negligible in many cases, may easily be estimated roughly by simple calculations, as has been done, for example, in the first chapter for the heat losses and the dissipation of the stirrer power.

In comparison to the heat transfer across the wall from the hot to the cold fluid, *longitudinal conduction* in the fluid may be neglected as long as

$$\boxed{Nu \ll (Re\,Pr)^2} \tag{2.221}$$

remains valid, as can be easily shown from an analysis accounting for longitudinal heat fluxes (problem!). The inequality of eq. (2.221) is valid for nearly all industrial heat exchangers. In laminar duct flow with Nu_∞ = const. and low Peclet numbers, the longitudinal conduction in the fluid can no longer be neglected. Especially for liquid metals ($Pr \ll 1$), this assumption, otherwise quite justified, does not hold. Further, and sometimes much more seriously, the restrictions of the validity of our analyses may be seen in the idealizing assumptions on the various degrees of longitudinal and lateral mixing of flow configurations. Real situations can be approximated only by these idealizing limiting cases—from "stirred tank" to "plug flow."

Besides the idealizing assumptions with respect to lateral and longitudinal mixing, it is frequently taken that the fluid flow is uniformly distributed. The effect of maldistribution can not always be completely avoided, especially in parallel flow channels (tube bundles, plate packs). It can, in fact, be surprisingly large and extremely unfavorable to the performance of exchangers, especially at high NTU [S4, M2]. In shell-and-tube heat exchangers, the effect of maldistribution is compounded by the effects of bypass flows. The usual design procedures these days choose the easier, though approximate, route of accounting for these effects by modifying the heat transfer coefficient [H3, V1]. This is not a logical approach as the effects ought to be accounted for in the balances, not in kinetic parameters. It is, nevertheless, better than completely ignoring these effects, so long as it is borne in mind that this approach is a temporary expedient. If theoretically expected efficiencies are to be achieved in practice, considerable attention has to be paid in the design and manufacture of heat exchangers, to ensure that the flow is distributed evenly among the parallel channels.

When numerous parallel channels are connected to common distributing and collecting ducts (Fig. 2.55), the usual approach for achieving uniform flow distribution among the channels has been to choose much larger cross-sectional areas for the distributor and collector ducts than those of the connecting channels. Recently, the

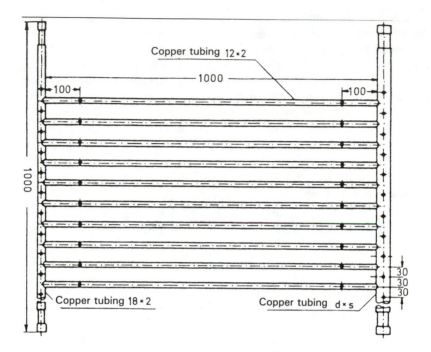

Figure 2.55 Flow distribution into *n* parallel channels.

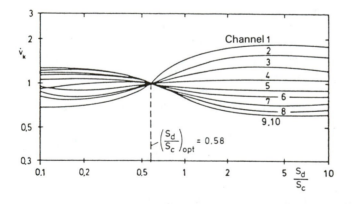

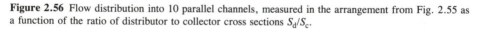

Figure 2.56 Flow distribution into 10 parallel channels, measured in the arrangement from Fig. 2.55 as a function of the ratio of distributor to collector cross sections S_d/S_c.

analysis of the manifold problem by Bassiouny and Martin [B2] revealed an alternative solution for nearly perfect uniform flow distribution in parallel channels that does not require large distributor and collector cross-sections [B1]. Only the ratio of distributor to collector cross-sections ($\neq 1$) needs to be chosen properly. Friction in the distributing and collecting ducts is neglected in the theory [B2] compared to the change of momentum arising from the withdrawal of mass into the parallel channels. A completely uniform flow distribution is possible if, and only if, the cross-sections of the distributor and the collector, S_d and S_c, are in the ratio:

$$\left(\frac{S_d}{S_c}\right)_{opt} = \left(\frac{2-\beta_d}{2-\beta_c}\right)^{1/2}$$

(2.222)

The quantities β_d and β_c are velocity ratios, defined as the local velocity components parallel to the distributor (collector) axis measured at the entry to (exit from) the parallel flow channels as related to the average flow velocity in the distributor (collector) cross-section at the location of the branch. These quantities, in the form ($2-\beta$), appear in the momentum balances written for each junction of the ducts and the branching channels. According to literature data and from Bassiouny's measurements [B1], the values of β_d are close to unity (actually somewhat larger, ≈ 1.3) while values of β_c around zero (or even slightly negative, ≈ -0.05) are found on the collector side. From this, optimal ratios of cross-sectional area of distributor and collector are found which significantly differ from unity:

$$\left(\frac{S_d}{S_c}\right)_{opt} \approx 0.58\ldots0.7$$

(2.223)

Experiments have shown that uniform distribution can indeed be achieved with (S_d/S_c) ≈ 0.58 for an arrangement of ten parallel tubes between distributor and collector tubes of only slightly larger size (Fig. 2.56). The diameter of the collector tube ought to be chosen about 30% larger than that of the distributor tube, in order to obtain uniform distribution among the parallel channels [$(1/0.58)^{1/2} \approx 1.3$].

In the following chapter, some examples will be given on how to treat the thermal and hydraulic design of different types of heat exchangers. In certain examples, not all the premises of our analyses in chapters 1 and 2 will hold good. This is mostly the case if the fundamental heat transfer process in an apparatus is coupled with other phenomena such as mass transfer, evaporation, and condensation.

THREE

EXAMPLES IN HEAT EXCHANGER DESIGN

1 WHAT IS THE OPTIMAL FLOW VELOCITY? EXAMPLE: DOUBLE-PIPE HEAT EXCHANGER

1.1 Problem Statement

In the following example, the design of an apparatus for a certain task will be demonstrated under relatively simple conditions, the data being set out below:

- In a continuous operation, 10,000 kg/h of aqueous leach (lixivium) are to be cooled from 60 °C to 20 °C. Cooling water at 10 °C is available, also at 10,000 kg/h.

1.2 Check for Feasibility, Minimum Heat Transfer Area Required

From the problem statement, the required efficiency (in other words, normalized temperature change) can be calculated immediately (subscript 1, leach; 2 cooling water):

115

$$\varepsilon_1 = \frac{T_1' - T_1''}{T_1' - T_2'}$$

$$\varepsilon_1 = \frac{60 - 20}{60 - 10} = \frac{4}{5} = 0.8$$

(3.1)

The aqueous leach and cooling water have roughly the same physical properties: as they also have the same flow rates, one can put $R = (Mc_p)_1/(Mc_p)_2 = 1$. At equal flow capacities, the required efficiency of 80% can only be reached by ideal counter flow, ideal crossflow, or countercurrent cascades (see Fig. 2.8 and Fig. 2.11). For ideal counterflow at equal flow capacities ($C = -1$),

$$\varepsilon = \frac{N}{1 + N} \qquad [\varphi(0) = 1]$$

(3.2)

Therefore,

$$N = \frac{\varepsilon}{1 - \varepsilon}$$

$$N = \frac{0.8}{0.2} = 4$$

(3.3)

The minimum required number of transfer units is, thus, 4 and the minimum required surface area is

$$\boxed{A_{req} = N_{min} \frac{\dot{M} c_p}{k}}$$

(3.4)

1.3 Physical Properties, Heat Duty

The average temperature is $T_{m1} = 40\,°C$ on the leach side, and $T_{m2} = 30\,°C$ on the cooling water side. For an approximate calculation, one may use the properties of pure water at $35\,°C$ (on both sides):

specific heat capacity	$c_{p1,2}$	$= 4.2 \cdot 10^3$ J/(kg K)
density	$\rho_{1,2}$	$= 1000$ kg/m^3
conductivity	$\lambda_{1,2}$	$= 0.62$ W/(Km)
viscosity	$\eta_{1,2}$	$= 720 \cdot 10^{-6}$ Pa s
Prandtl number ($Pr = \eta c_p/\lambda$)	Pr	$= 4.9$
The heat duty to be transferred is	$\dot{Q} = \dot{M} c_p(T_1' - T_1'')$	$= 467$ kW

1.4 Choice of Dimensions of One Heat Exchanger Element

We choose a double-pipe (see Fig. 3.1), the center tube made from stainless steel (leach) with a heat conductivity of $\lambda = 12$ W/(K m). First, an order of magnitude of the area can be found from eq. (3.4) using an estimated value of k (water-to-water, double-pipe) ≈ 1000 W/(m^2 K) to be $A_{req}^{(0)} = 4(10000/3600)(4.2 \cdot 10^3/1000)$ m$^2 = 47$ m^2.
    Choosing conventional tubing dimensions, as, e.g.,

central tube	25 × 2.5 [mm × mm]	$(d_o \times s)_T$
outer tube (shell)	38 × 3.5 [mm × mm]	$(d_o \times s)_S$

and a length of $L = 6$ m, then the transfer surface area of one element according to Fig. 3.1, the outer diameter of the central tube being chosen as the reference diameter, is

$$A_{Element} = \pi d_{oT} L = \pi 25 \cdot 10^{-3} \text{ m} \cdot 6 \text{ m} = 0.4712 \text{ m}^2 \qquad (3.5)$$

About 100 such elements would, therefore, be needed for the task at hand. These elements may be connected partly in parallel and partly in series. By means of the flow cross-sections of one element and the number of elements connected in parallel, one can change the flow velocity w and, consequently, the heat transfer coefficients.
    The flow cross-sections of a heat exchanger element are

tubeside cross-section	$S_T = \pi/4(d_o - 2s)_T^2 = 3.142$ cm^2
shellside cross-section	$S_S = \pi/4[(d_o - 2s)_S^2 - d_{oT}^2] = 2.639$ cm^2

The cross-sections of central tube and shell side (annulus) are roughly equal ($S_T/S_S = 1.190$). The flow velocity is obtained from continuity with the number n_p of parallel elements:

$$w = \frac{\dot{M}}{\varrho n_p S} \qquad (3.6)$$

with $n_p = 1$ (all the elements connected in series), we would get a maximum flow velocity in the annulus of about $w_{max} \approx 10$ m/s. With $n_p = 100$ (all the elements in

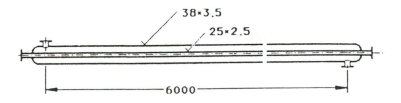

Figure 3.1 Double-pipe heat exchanger element.

in parallel), $w_{min} \approx 0.1$ m/s would be obtained. Which arrangement, i.e., which flow velocity, would be most favorable?

1.5 The Economically Optimal Flow Velocity

It is well known that heat transfer coefficients α for flow through ducts increase with increasing flow velocity w, with heat conductivity λ, density ρ, and specific heat capacity c_p of the fluid, while they decrease with increasing viscosity η of the fluid, and with the diameter d as well as the length L of the duct. The exact relationships may be found from the standard equations [H3, V1] valid in the laminar or turbulent regimes, respectively, at specified thermal boundary conditions (problem!).

Once the physical properties (λ, ρ, c_p, η) and the dimensions of the heat exchanger element (d, L) are fixed, the choice of the highest possible flow velocity w leads to the smallest transfer surface area, i.e., to the smallest number of exchanger elements required. The *investment costs* for the apparatus will, therefore, diminish with increasing flow velocity. High flow velocities, however, cannot be obtained free of charge. The fluids have to be transported through the apparatus by means of pumps or compressors. The pumping power $\dot{W}_P$ for given mass flow rate and density is proportional to the pressure drop across all the elements connected in series:

$$\dot{W}_P = \dot{M}\frac{\Delta p}{\varrho} \tag{3.7}$$

From the equation

$$\Delta p = \xi(Re)\frac{\varrho}{2}w^2\frac{n_s L}{d} \tag{3.8}$$

where $n_s L$ is the total length of all elements placed in series, one can recognize that the pumping power and, therefore, the *cost of operation* will strongly increase with flow velocity. An economically optimal flow velocity w_{opt} will, therefore, be found, if the sum of the investment costs, decreasing with w, and the costs of operation, rising with w, reaches a minimum.

Such an economic optimization will be demonstrated in principle below using rather simple assumptions.

1.6 Calculation of Required Transfer Surface Area

First, as a base value, a velocity of $w_T \approx 1$ m/s in the inner tube is chosen. From eq. (3.6), it is seen that $n_p = 9$ tubes connected in parallel will suffice. The flow velocities are then

$$w_T = \left(\frac{\dot{M}}{\varrho n_p S}\right)_T = 0.982 \text{ m/s}$$

and

$$w_S = \left(\frac{\dot{M}}{\varrho n_p S} \right)_S = 1.17 \text{ m/s}$$

For the inner tube, this results in a Reynolds number of

$$Re_T = \frac{\varrho w_T d_{hT}}{\eta} = 27\,300 \qquad (\rightarrow \text{turbulent})$$

With

$$\frac{d_{hT}}{L} = \frac{20 \cdot 10^{-3}}{6}$$

and

$$Pr = 4.9$$

one obtains from eqs. (1.124) and (1.140)

$$\boxed{Nu_T = 172 \qquad \xi_T = 0.0242 \qquad \alpha_i = 5\,330 \text{ W/m}^2 \text{ K}}$$

On the shell side (in the annulus) with

$$d_{hS} = (d_o - 2s)_S - d_{oK} = 6 \text{ mm}$$

$$Re_S = \frac{\varrho w_S d_{hS}}{\eta} = 9\,750 \quad (\rightarrow \text{ turbulent})$$

$$\frac{d_{hS}}{L} = 10^{-3}$$

$$K = \frac{d_{oT}}{(d_o - 2s)_S} = \frac{25}{31}$$

$$f_i = 0.86\, K^{-0.16} = 0.8901$$

$$Pr = 4.9$$

one finds correspondingly:

$$\boxed{Nu_S = 60.9 \qquad \xi_S = 0.0317 \qquad \alpha_o = 6290 \text{ W/m}^2\text{K}}$$

The overall heat transfer coefficient, defined for the outer surface of the central tube (A_{oT}) follows from

$$\boxed{\frac{1}{k} = \frac{A_{oT}}{\alpha_i A_{iT}} + \frac{s_T A_{oT}}{\lambda_T A_{mT}} + \frac{1}{\alpha_o} + R_f} \tag{3.9}$$

Herein is

$$\frac{A_{oT}}{A_{iT}} = \frac{d_{oT}}{(d_o - 2s)_T} = \frac{25}{30} = 1.250$$

$$A_m = \frac{A_o - A_i}{\ln(A_o/A_i)} \rightarrow \frac{A_{oT}}{A_{mT}} = \frac{25}{5}\ln\frac{25}{20} = 1.116$$

R_f is an additional heat transfer resistance, accounting for fouling of the transfer surfaces during operation. Empirical values for this may be found in the literature depending on the kind of fluids and the conditions of operation [H3, V1]. The component resistances in eq. (3.9) are best expressed in [m^2 K/kW]:

$$\frac{1}{k} = (0.235 + 0.232 + 0.159 + 0.1)\ \text{m}^2\text{K/kW}$$
$$\text{inside}\quad \text{wall}\quad \text{outside}\quad R_f$$

$$\frac{1}{k} = 0.726\ \text{m}^2\text{K/kW} \longrightarrow \boxed{k = 1\,378\ \text{W/m}^2\text{K}}$$

In this case, obviously all the resistances are of the same order of magnitude. The fouling resistance has been chosen to be $R_f = 0.1$ m^2 K/kW. From eq. (3.4), the required transfer surface area becomes

$$A_{req} = 33.87\ \text{m}^2 \qquad \frac{A_{req}}{A_{Element}} = 71.9$$

i.e.,

$$\boxed{n = n_s n_p} = 8 \cdot 9 = 72$$

and

$$\boxed{A = 72\, A_{Element} = 33.9\ \text{m}^2}$$

1.7 Calculation of Pumping Power

The calculation of pumping powers from eqs. (3.7) and (3.8) is now relatively simple, as the friction factors ξ had already been calculated to find the heat transfer coefficients. In addition to the straight lengths of tubing, pressure losses in the 180° bends on the tube side and in the sharp-edged T-junctions on the shell side, as in Fig. 3.2, have to be accounted for.

Tube side

$$\Delta p_T = n_s \left(\xi_T \frac{L}{d_{hT}} + \xi_{180°\ \text{Bend}} \right) \frac{\rho}{2} w_T^2 \tag{3.10}$$

The loss coefficient for smooth 180° bends depends on the radius of curvature r and

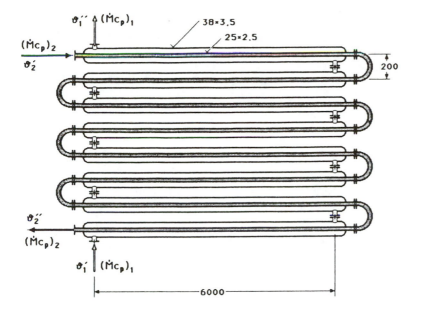

Figure 3.2 Double-pipe heat exchanger, series connection of double-pipes.

weakly on the Reynolds number. From information given in HEDH 2.22–16, one finds

$$\xi_{180° \text{ Bend}} = 1.38 \cdot 0.216 \left(\frac{r}{d_i/2} \right)^{0.84} \left[0.95 + 17.2 \left(\frac{r}{d_i/2} \right)^{-1.96} \right] Re^{-0.17}$$

which can be written more concisely as

$$\xi_{180° \text{ Bend}} = 0.283 Re^{-0.17} \left[\left(\frac{2r}{d_i} \right)^{0.84} + 18.1 \left(\frac{2r}{d_i} \right)^{-1.12} \right] \qquad (3.11)$$

With $(2r/d_i) = 200/20 = 10$, we obtain

$$\xi_{180° \text{ Bend}} \approx 2.35 Re^{-0.17} \qquad (3.12)$$

with

$$Re_T = 27\ 300 \rightarrow \xi_{180° \text{ Bend}} = 0.41$$

$$\Delta p_T = 8(7.26 + 0.41) \frac{1000}{2} (0.982)^2 \text{ Pa}$$

and

$$\Delta p_T = 29\ 600\ \text{Pa}\ (= 0.296\ \text{bar})$$

With this, the pumping power becomes $\dot{W}_{PT} = 82.2$ W. Dividing this by the pump efficiency, taken here as 70%, the electric power required to drive the pump works out to $\dot{W}_{PT,el} = 117$ W.

Shell side

$$\Delta p_S = n_s \left(\xi_S \frac{L}{d_{hS}} + \xi_{180°\ \text{sharp}} \right) \frac{\rho}{2} w_S^2 \tag{3.13}$$

The loss coefficient for a sharp bend of $2 \times 90°$ in a T-junction may be taken from suitable diagrams in HEDH, VDI–WA or other handbooks. One, thus, finds

$$\xi_{180°\ \text{sharp}} \approx 1.3 + 1.05 = 2.35 \tag{3.14}$$

i.e.,

$$\Delta p_S = (31.7 + 2.35) \frac{1000}{2} (1.17)^2\ \text{Pa}$$

$$\Delta p_S = 186\ 000\ \text{Pa}\ (= 1.86\ \text{bar}).$$

The pumping power is, therefore,

$$\dot{W}_{PS} = 518\ \text{W}$$

and the electric driving power becomes

$$\boxed{\dot{W}_{PS,el} = 740\ \text{W}}$$

1.8 Calculation of Costs

Costs of investment

Heat exchanger elements such as the one shown in Fig. 3.1 cost 500 $ each including all fittings (screws, gaskets, etc.). With an amortization of 10%/a, this amounts to an investment of 50 $/a. The investment costs required for n elements are, therefore,

$$\boxed{C_1 = n \cdot 50\ \$/a} \tag{3.15}$$

At $w_T = 0.982$ m/s, we have $n = 72$

$$\rightarrow \boxed{C_1 = 3\ 600\ \$/a}$$

Costs of operation

If the plant is to be in operation for $\tau = 6000$ h/a (out of the maximum possible 8,760 hours per year) and if the price of electrical energy would be $k_{el} = 0.15$ \$/kWh, the operation would lead to the costs

$$C_O = \tau k_{el}(\dot{W}_{PT,el} + \dot{W}_{PS,el}) \tag{3.16}$$

$$C_O = 900 \dot{W}_{P,el} \ \$/(\text{akW})$$

and with

$$\dot{W}_{P,el} = (117 + 740)10^{-3} \ \text{kW} \qquad \rightarrow \boxed{C_O = 771 \ \$/\text{a}}$$

Total costs

$$C = C_I + C_O \qquad \rightarrow \boxed{C = 4\ 371 \ \$/\text{a}}$$

1.9 Optimization, Discussion of Results

The calculations from sections 1.6 (A, n), 1.7 $(\dot{W}_P, \dot{W}_{P,el})$, and 1.8 (C_I, C_O, C) have to be repeated now with other velocities, or numbers of parallel elements n_p (problem!). For these repeat calculations, it would be best to proceed by halving and doubling the velocity used initially:

a. $\qquad n_p = 4 \qquad \rightarrow \qquad w_T = 2.21$ m/s $\qquad w_S = 2.63$ m/s

b. $\qquad n_p = 18 \qquad \rightarrow \qquad w_T = 0.491$ m/s $\qquad w_S = 0.585$ m/s

One obtains:

a.	$n_s = 14$	$n = 56$	$C_I = 2\ 800$ \$/a
	$\Delta p_T = 217$ kPa	$\Delta p_S = 1\ 348$ kPa	
	$\dot{W}_{PT,el} = 863$ W	$\dot{W}_{PS,el} = 5\ 351$ W	$C_o = 5\ 539$ \$/a
			$C = 9\ 393$ \$/a
b.	$n_s = 6$	$n = 108$	$C_I = 5400$ \$/a
	$\Delta p_T = 6.61$ kPa	$\Delta p_S = 42.3$ kPa	
	$\dot{W}_{PT,el} = 26.2$ W	$\dot{W}_{PC,el} = 168$ W	$C_o = 175$ \$/a
			$C = 5\ 575$ \$/a

or all together:

$w_T/(m/s)$	n_p	n_s	n	$C_I/(\$/a)$	$C_o/(\$/a)$	$C/(\$/a)$
0.49	18	6	108	5400	175	5575
0.98	9	8	72	3600	771	4371
2.2	4	14	56	2800	5593	8393

In Fig. 3.3, individual and total costs are plotted versus the velocity in the inner tube. One can see a minimum in the total cost at a velocity of somewhat above 1 m/s. Precise calculation would show that the most favorable apparatus has $n_p \cdot n_s = 8 \cdot 8 = 64$ elements. With a tube pitch of 200 mm, the envelope dimensions would be roughly $(1.6 \times 1.6 \times 6)$ m^3.

In order to demonstrate the principle, the calculations of costs were kept as simple as possible. In practice, many more individual costs will have to be accounted for. So the result obtained here is of a more-or-less qualitative nature.

The pumps will have to overcome not only the flow resistances in the heat exchanger but also in the connecting pipes and other components of the plant. An economic optimization, then, has to be extended to the whole plant. Heat exchanger design will, therefore, usually be based on a given allowable pressure drop which has been specified by a plant designer for the entire heat exchanger flow circuit. The designer of the heat exchanger, then, has to use this given pressure drop as well as possible, to keep investment low.

The problem of fouling of heat transfer surfaces may also lead to the choice of high flow velocities. Conventionally used *flow velocities of liquids* in heat exchangers are in the range

$$\boxed{0.2 \text{ m/s} < w_1 < 2 \text{ m/s}} \tag{3.17}$$

and those *of gases* at atmospheric pressures are in the range

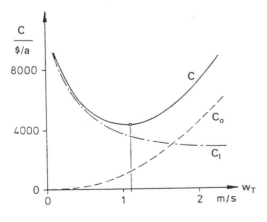

Figure 3.3 Investment-, operation- and total costs vs. flow velocity in inner tube.

$$\boxed{5 \text{ m/s} < w_\text{g} < 50 \text{ m/s}} \tag{3.18}$$

i.e., at velocities higher by a factor of around 25. This can be explained from the fact that the costs of investment ($C_1 \propto A_{\text{req}} \propto 1/k$) are much higher for gas-gas heat exchangers at the same thermal task and flow velocity than those for liquid-liquid heat exchangers ($k_1/k_\text{g} \propto 500$), while the costs of operation ($C_\text{O} \propto \dot{W}_\text{P} \propto \xi \rho w^3$) are much lower for gases than for liquids because of the much lower densities of gases. Therefore, the minimum in total cost shifts to much higher velocities. In specific applications, the given ranges of velocity may be exceeded on either side.

2 WHAT ARE THE OPTIMAL DIMENSIONS OF A HEAT EXCHANGER? EXAMPLE: SHELL-AND-TUBE APPARATUS

2.1 Problem Statement

In the preceding example, the question for an optimal flow velocity was treated for an apparatus with specified dimensions of one element. The result was an arrangement of 8 × 8 double pipes (eight assemblies in parallel, with each assembly having eight tubes and annuli in series) in an envelope of 1.6 × 1.6 × 6 m³. In place of individual shells, one could also have used one common large shell for all the inner tubes, i.e., a shell-and-tube (bundle) apparatus. At a given flowrate and pressure drop, one can realize the required total flow cross-section S by a few large or by many small tubes. Is there an optimum for the tube diameter d, and thus for the number of tubes in the bundle? To simplify the treatment of this question, we assume that the shell-side fluid evaporates at constant temperature and that the heat transfer resistance is on the tube side ($C = 0$, $k \approx \alpha_\text{i}$). One could imagine, for example, that air flows through the tubes and is to be cooled from T' to T''. On the shell side, a refrigerant may evaporate, as shown schematically in Fig. 3.4.

Because $k \approx \alpha_\text{i}$, the tube wall temperature T_w is constant and equal to the refrigerant saturation temperature corresponding to its pressure. With the duty to cool the given flow rate M of air from T' to T'', the efficiency ϵ and the required number of transfer units are fixed:

$$N_{\text{req}} = \frac{\varepsilon}{\Theta} = \frac{T' - T''}{\triangle T_{\text{LM}}} = \text{const} \tag{3.19}$$

2.2 Feasibility

The required NTU

$$N_{\text{req}} = \frac{kA}{\dot{M}c_\text{p}} \approx \frac{\alpha_\text{i}A}{\dot{M}c_\text{p}} \tag{3.20}$$

$$N_{\text{req}} = \frac{A}{S} \frac{Nu}{Re\,Pr}$$

can also be expressed in terms of the dimensionless numbers Nu and $Re \cdot Pr \cdot d/L = Gz$ for duct flow ($\dot{M} = \rho w S$, $A/S = 4\,L/d_h$). From this, the ratio of length to diameter of the tubes has to satisfy the condition

$$\frac{L}{d} = N_{\text{req}} \frac{Re\,Pr}{4N\,u} \tag{3.21}$$

In turbulent duct flow, the Nusselt number has only a weak dependence on L/d (see eq. [1.140]). In the turbulent range, $2300 \leq Re \leq 10^6$ and at fixed Prandtl number (say, $Pr = 0.7$, air) the factor $Re \cdot Pr/(4\,Nu)$ varies from 56 to 155, a less than threefold variation. Thus, once the thermal task, i.e., N_{req}, is specified, the ratio of length to diameter is more or less fixed in the turbulent range of operation.

Large NTU requires long tubes!

(See problem 2.8 and Fig. 2.12) This can also be seen from a graph of the function $Nu(Gz, Pr)$—a Nusselt–Graetz chart—as shown in Fig. 3.5 for $Pr = 0.7$. In that diagram, as favored by Schlünder [H3], the lines $N = $ const. from eq. (3.21) are straight lines (radii through the origin) with a slope of $N/4$. In a log-log plot, these are lines $\lg Nu = \lg (N/4) + \lg Gz$ with a slope of unity and an intercept $\lg (N/4)$. The curves calculated from eq. (1.140) for turbulent duct flow are shown for some values of L/d as a parameter. Over a wide range, they are nearly parallel to the lines of constant NTU.

A tube with a diameter of 100 mm and a length of 10 m does not transfer more heat than one with a diameter of 10 mm and a length of 1 m at the same throughput, though it has a hundredfold larger transfer surface area. The pressure drop in the smaller tube would be higher, however, by almost a factor of ten thousand! In the turbulent range, high efficiency ($\epsilon \rightarrow 1$, i.e., $N \gg 1$) can only be achieved with long

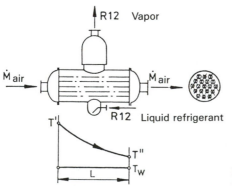

Figure 3.4 Cooling of air by vaporizing refrigerant in shell-and-tube heat exchanger.

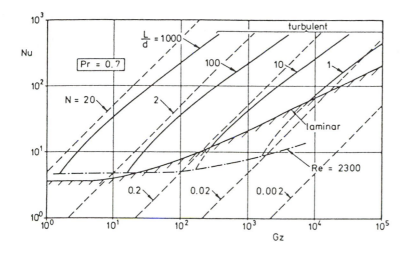

Figure 3.5 Nusselt-Graetz chart for tube flow, air ($Pr = 0.7$).

tubes (large L/d). Also shown in Fig. 3.5 is a single curve of $Nu(Gz)$ at a constant $Pr = 0.7$ for laminar flow. One may conclude from the figure that *it is possible to realize any required NTU with laminar flow too*. In this case, with a specified NTU, the value of $Re \cdot Pr \cdot d/L = Gz$ is also fixed immediately:

$$N = 4Nu(Gz)/Gz \tag{3.22}$$

$$Gz = \frac{\varrho c_p w d^2}{\lambda L} = \frac{4\dot{M}c_p}{\pi \lambda n L} \tag{3.23}$$

Here $\dot{M} = \varrho w n \pi d^2/4$ is the total mass flow through the n parallel tubes in the bundle. Now the thermal problem (flow rate, physical properties, change in temperature, i.e., NTU and, consequently, also Gz given) can be solved under the condition

$$\boxed{n L = c_T} = \text{const} \tag{3.24}$$

with bundles of various numbers n and lengths L of the tubes if only the product $n \cdot L$ is kept constant. The pressure drop and, thus, the pumping power remain constant, according to the Hagen–Poiseuille law (see eq. [1.114] with $K = 0$ and $M \rightarrow M/n$) if the condition

$$\boxed{\frac{L}{nd^4} = c_H} = \text{const} \tag{3.25}$$

is also fulfilled. Both the thermal condition ($c_T = $ const) and the hydrodynamic condition ($c_H = $ const) may be simultaneously fulfilled by keeping the ratio

$$\frac{L}{d^2} = (c_T c_H)^{1/2} \tag{3.26}$$

constant. With that is also fixed the relation

$$n\,d^2 = \left(\frac{c_T}{c_H}\right)^{1/2} = \text{const} \tag{3.27}$$

i.e., the tube-side total cross-section $S = n(\pi/4)d^2$ and, thus, the flow velocity are kept constant. Therefore, the transfer surface area

$$A = n\pi dL \tag{3.28}$$

can be expressed as a function of the number of parallel tubes alone:

$$A = \left(\pi c_T^{5/4} c_H^{-1/4}\right) n^{-1/2} \tag{3.29}$$

The transfer surface area of a tube bundle apparatus can be arbitrarily reduced, inversely proportional to the square root of the number of tubes at the same performance and the same pressure drop within the range of laminar flow!

As an example, Fig. 3.6 shows three tube bundles the dimensions of which are related as

$$
\begin{array}{ccccccccc}
n_1 & : & n_2 & : & n_3 & = & 1 & : & 4 & : & 16, \\
L_1 & : & L_2 & : & L_3 & = & 16 & : & 4 & : & 1, \\
d_1 & : & d_2 & : & d_3 & = & 4 & : & 2 & : & 1 & \text{and} \\
A_1 & : & A_2 & : & A_3 & = & 4 & : & 2 & : & 1
\end{array}
$$

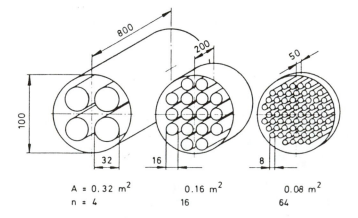

Figure 3.6 Three shell-and-tube heat exchangers with the same pressure drop and the same transfer performance.

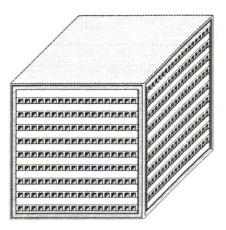

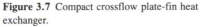

Figure 3.7 Compact crossflow plate-fin heat exchanger.

from eqs. (3.24), (3.27), and (3.29). All three bundles can transfer the same heat at laminar flow in the tubes at the same pressure drop.

2.3 Optimal Dimensions of a Heat Exchanger

From the considerations of the preceding section, the costs of operation, resulting from pumping power, may be kept as low as desired if only the flow cross-section $S = n(\pi/4)d^2$ is chosen large enough. For laminar flow, the pumping power $\dot{W}_P = \Delta p \cdot \dot{V}$ is related to the constant c_H in eq. (3.25):

$$c_H = \frac{\pi}{128} \frac{\dot{W}_P}{\eta \dot{V}^2} \qquad (3.30)$$

Now the transfer surface area from eq. (3.29), $A \propto 1/\sqrt{n}$, can be made arbitrarily small in principle, if only more and more ever smaller tubes (with $d \propto 1/\sqrt{n}$ and $L \propto 1/n$) are arranged in the same total cross-section.

The optimal dimensions of a heat exchanger would, then, be those of a disc with a huge number of tiny tubelets, the diameters and length of which could tend to zero with increasing number n. *Volume and surface area of that optimal heat exchanger would then also tend to zero!*

What are the practical limits of such a miniaturization of heat exchangers? If the length L, according to eq. (3.24), is reduced more and more, inversely proportional to the increasing number of tubes n, then, at $n \to \infty$ and $L \to 0$, there will be no space left for the other fluid (e.g., the evaporating liquid here) to enter and leave the shell side at the rate required by the heat duty. Thus, we end up with the design problem to distribute two fluids in one plane into two spaces enclosing each other in extremely fine dispersion.

The miniaturization of the flow channels is also limited by fouling, which can not be avoided in every case. Nevertheless, diameters of flow channels in the range of millimeters can certainly be realized in practice. The discs of the rotating regenerators consisting of a solid matrix with fine parallel channels (see Fig. 2.45) obviously come close to the optimal geometry. The distribution problem is solved in this case by pumping both fluids through the same channels successively.

Even without rotating storage masses one can build and operate very compact exchangers with channel diameters below one millimeter [K1, C4] provided that the operation conditions are favorable (very clean fluids required).

Only recently an extremely compact plate-fin crossflow heat exchanger with rectangular flow channels of 90×95 $(\mu m)^2$ and wall and fin thicknesses of 18 μm has been built (see Fig. 3.7). The manufacturing process developed at the nuclear research center Karlsruhe would allow for even smaller dimensions [B4]. In a cube of edge 1 cm, about 4000 channels can be placed on the side of each fluid with a total transfer surface area of 150 cm^2 (i.e., a specific surface area of $a_v = 15 \times 10^3$ m^2/ m^3, in the same order of magnitude as that of the human lungs). Operated with water as the fluid on both sides and flowrates of about 6 liters/min, volumetric overall heat transfer coefficients (ka_v) as high as 200 MW/(m^3 K) have been achieved [B4]! For the operation of such *micro heat exchangers,* a high purity of the fluids is required, which may be possible in some special applications with closed circuits. It is difficult, however, to build such exchangers for larger flowrates. The plate heat exchangers described in chapter 2, section 4, with their gap widths in the order of a few millimeters are certainly a reasonable compromise in this respect.

3 HEAT EXCHANGERS WITH COMBINED FLOW CONFIGURATIONS EXAMPLE: PLATE HEAT EXCHANGERS AND SHELL-AND-TUBE HEAT EXCHANGERS

3.1 Problem Statement

Wastewater from the bottle cleaning plant of a brewery with a flow rate of 36 m^3/h and a temperature of 80 °C is to be used to preheat 90 m^3/h of freshwater at 20 °C to the highest possible final temperature. The wastewater has to be cooled down to 30 °C to be led into the purification plant.

3.2 Check for Feasibility, Performance

The normalized change in temperature of the wastewater is

$$\varepsilon_1 = \frac{T_1' - T_1''}{T_1' - T_2'} \tag{3.31}$$

$$\varepsilon_1 = \frac{50}{60} = 0.833$$

At $(\rho c_p)_1 = (\rho c_p)_2 = 4.19 \cdot 10^6$ J/(m^3 K), the capacity ratio becomes

$$R = \frac{(\varrho c_p)_1 \dot{V}_1}{(\varrho c_p)_2 \dot{V}_2} \tag{3.32}$$

$$R = \frac{36}{90} = 0.4$$

and so we get $\epsilon_2 = 0.4 \cdot \epsilon_1 = 0.333$. The 2.5-fold larger freshwater stream can, therefore, be preheated to $T''_2 = 40\,°C$. The heat duty is

$$\dot{Q} = (\varrho c_p)_1 \, \dot{V}_1 (T'_1 - T''_1)$$
$$\dot{Q} = 4.19 \cdot 10^6 \cdot 0.010 \cdot 50 \text{ W} = 2.095 \text{ MW} \tag{3.33}$$

With pure counterflow, this would require a minimum number of transfer units $N_{1,\min}$ of

$$N_{1,\min} = \frac{\varepsilon_1}{\Theta_{LM}} \tag{3.34}$$

From

$$\Theta_{LM} = \frac{(1 - \varepsilon_2) - (1 - \varepsilon_1)}{\ln[(1 - \varepsilon_2)/(1 - \varepsilon_1)]}$$

$$\Theta_{LM} = \frac{1}{2 \ln 4} = 0.361$$

(see Table 2.1) one finds:

$$\boxed{N_{1,\min} = 2.31}$$

The transfer surface area required, thus, becomes

$$\boxed{A_{req,min} = N_{1,\min} \frac{(\varrho c_p \dot{V})_1}{k}} \tag{3.35}$$

3.3 Choice of Type and Determination of Size

3.3.1 Plate heat exchanger. It makes sense to choose a relatively compact type like the plate heat exchanger (chapter 2, section 4) since corrosion resistant, hence expensive, materials (stainless steel) are necessary because of the aggressive cleansing agents in the wastewater.

With an estimated overall heat transfer coefficient of $k^{(0)} = 2\,000$ [W/(m² K)], we find a minimum surface area required from eq. (3.35) of

$$A_{req,min}^{(0)} = 2.31 \frac{4.19 \cdot 10^6 \cdot 0.01}{2\,000} \text{ m}^2 \approx 48 \text{ m}^2$$

From the plate sizes available, the plate chosen should yield the smallest possible ratio of envelope surface to the volume of the stack built up from the plates. The plate pack has the dimensions $(B \cdot L \cdot D) = V$. The ratio of outer surface of the packet

$$A_{\text{envelope}} = 2(BL + BD + DL)$$

to its volume is

$$\left(\frac{A_{\text{envelope}}}{V}\right)_{\text{packet}} = 2\left(\frac{1}{B} + \frac{1}{L} + \frac{1}{D}\right) \tag{3.36}$$

The thickness of the plate pack, termed D, follows from the number of plates required $n_p = A_{\text{req}}/A_P$ and the sum of gap width b and wall thickness s:

$$D = n_P(b + s) \tag{3.37}$$

From Table 3.1, showing typical dimensions of pressed plates [P2], one can find the lowest values of $(A_{\text{envelope}}/V)_{\text{pack}}$ from eq. (3.36) for plates of the sizes 4 ($n_p = 150$), 5 ($n_p = 100$), and 6 ($n_p = 80$).

The flow velocity should be chosen in the range $w = 0.5$ to 0.8 m/s for plates with a "hard" pattern. In the case of the larger stream of freshwater, this requires a flow cross-section of

$$S_2 = \frac{\dot{V}_2}{w_2} \tag{3.38}$$

With $S = n_c Bb$ and size 5 plates (from Table 3.1), a number of parallel channels of $n_c = 20 \ldots 31$ can be found. We choose $n_c = 25$ channels in two passes for the

Table 3.1 Typical dimensions of pressed plates.

Size	Width B/m	Length L/m	Plate surface A_P/m^2	Port diameter D_P/m
1	0.15	0.70	0.11	0.08
2	0.30	0.50	0.15	0.12
3	0.30	0.80	0.24	0.12
4	0.40	0.80	0.32	0.15
5	0.40	1.20	0.48	0.15
6	0.60	1.00	0.60	0.15
7	0.60	1.40	0.84	0.15
8	0.80	1.00	0.80	0.18
9	0.80	1.40	1.12	0.18
10	0.80	1.60	1.28	0.18

Gap width $b = 4$ mm, wall thickness $s = 0.65$ mm.

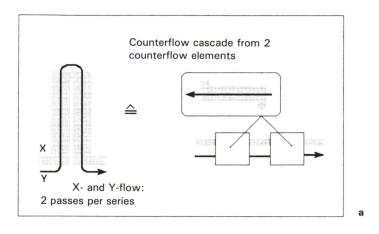

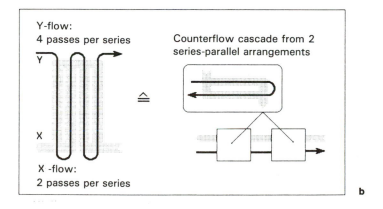

Figure 3.8 Flow configuration in a plate exchanger, example: 2 and 4 passes.

freshwater side, i.e., 50 channels, and, therefore, $n_p = 2 \cdot 50 + 1 = 101$ plates of size 5. Then, the flow velocity on the freshwater side becomes

$$w_2 = \frac{90}{3\,600 \cdot 25 \cdot 0.4 \cdot 4 \cdot 10^{-3}} \text{ m/s}$$
$$w_2 = 0.625 \text{ m/s}$$

If we choose two passes on the wastewater side also, we would get a velocity of only

$$w_1 = 0.250 \text{ m/s}$$

A higher heat transfer coefficient could be gotten by choosing four passes on the wastewater side and, thus, $w_1 = 0.5$ m/s.

In the case of *two passes* on

a. **both sides,** we get a countercurrent cascade of two counterflow elements (see Fig. 3.8*a*)
b. **the freshwater side,** and **four passes on the wastewater side,** the flow configuration could be described as a countercurrent cascade of two 1 × 2 series-parallel arrangements (see Fig. 3.8*b*).

For an estimate, let us assume that, at $w = 0.5$ m/s, a heat transfer coefficient of $\alpha \approx 10{,}000$ [W/(m^2 K)] would be reached, and that $\alpha \propto w^{0.75}$ is valid. With fouling resistances

$$R_{f1} = 0.15 \text{ m}^2\text{K/kW} \quad \text{(wastewater)}$$

$$R_{f2} = 0.10 \text{ m}^2\text{K/kW} \quad \text{(freshwater)}$$

and a wall resistance

$$\frac{s}{\lambda_w} = \frac{0.65}{14.5} \text{ m}^2\text{K/kW}$$

we can find the overall heat transfer coefficient in case (a) to be around:

$$\left(\frac{1}{k}\right)_a = (0.0846 + 0.2948 + 0.1682) \text{ m}^2\text{K/kW}$$

$$\longrightarrow \quad k_a \approx 1\,830 \text{ W/(m}^2\text{K)}$$

and in case (b) to be:

$$\left(\frac{1}{k}\right)_b = (0.0846 + 0.2948 + 0.100) \text{ m}^2\text{K/kW}$$

$$\longrightarrow \quad k_b \approx 2\,090 \text{ W/m}^2\text{K}$$

From this $k_b/k_a \approx 1.14$. The countercurrent cascade of two series-parallel arrangements (Fig. 3.8*b*), however, has a lower mean temperature difference $\Delta T_M = (T_1' - T_2')\cdot\Theta$ than the countercurrent cascade of two counterflow elements. The latter is, in fact, a pure counterflow configuration (Fig. 3.8*a*). Recalling eq. (2.82), at $R \neq 1$, the efficiency of one element of a cascade can be found from

$$\frac{1 - R\varepsilon_j}{1 - \varepsilon_j} = \left(\frac{1 - R\varepsilon}{1 - \varepsilon}\right)^{1/J} = \left(\frac{\Delta\vartheta_0}{\Delta\vartheta_1}\right)_j \qquad (3.39)$$

$$\varepsilon := \varepsilon_1 = \frac{5}{6}, \quad R = 0.4, \quad J = 2$$

$$\longrightarrow \quad \boxed{\varepsilon_j = 0.625}$$

$$\longrightarrow \quad \Theta_{LM} = \frac{(1-R)\varepsilon_j}{\ln(\Delta\vartheta_0/\Delta\vartheta_1)_j} = 0.5410$$

From eq. (2.209), with $Y = N_1/2$, $X = 0.4 \cdot Y$, we get for the (1×2) series-parallel arrangement

$$\varepsilon_{j(1\times2)} = 1 - (1-\varepsilon_p)(1-\varepsilon_c)$$

$$\varepsilon_p = \frac{Y/2}{\varphi[(Y/2)+X]}, \qquad \varepsilon_c = \frac{Y/2}{\varphi[(Y/2)-X]+X} \qquad (3.40)$$

$$\Theta_{j(1\times2)} = \frac{\varepsilon_{j(1\times2)}}{Y} \qquad (3.41)$$

$$F_{(1\times2)} = \frac{\Theta_{j(1\times2)}}{\Theta_{j,LM}} \qquad (3.42)$$

N_1 has to be somewhat greater than 2.31, the value of $N_{1,min}$ calculated by way of feasibility check in section 3.2.

N_1	2.31	2.4	2.5
$Y/2$	0.578	0.600	0.625
X	0.462	0.480	0.500
ε_p	0.359	0.367	0.375
ε_c	0.380	0.389	0.400
ε_j	0.603	0.613	0.625
$\Theta_{j(1\times2)}$	0.522	0.511	0.500
$F_{(1\times2)}$	0.964	0.945	0.924

The mean temperature difference in case (b) will be reduced by the factor $F = 0.924$ compared to pure counterflow (case a). As it has a 14% higher overall heat transfer coefficient, option b is more favorable ($0.924 \times 1.14 = 1.053$), as long as the slightly higher costs of operation are not important.

The transfer surface area required is

$$A_{req} = N_1 \frac{(\varrho c_p \dot{V})_1}{k} = \frac{2.5 \cdot 4.19 \cdot 10^6 \cdot 0.01}{2\,090} \text{ m}^2 \approx 50 \text{ m}^2$$

From this, the number of plates and the velocities now become

$$n_P = 105, \quad n_C = 52, \quad n_{C,parallel,1} = 13, \quad n_{C,parallel,2} = 26$$

$$w_1 = 0.481 \text{ m/s}, \quad w_2 = 0.601 \text{ m/s}$$

A more accurate recalculation can be carried out if equations for Nu, such as eq. (2.122) for example, are available for the actually used plates. Often the effort to do such a recalculation may not be justified with respect to the uncertainty of the fouling resistances. The plate pack is $1.20 \times 0.40 \times 0.49$ m^3 (vertical height $L \times$ width of the plates $B \times$ thickness of the plate pack or depth D). To mount and remove the plates, an additional depth of the frame of about 0.5 m will be needed, so that the whole apparatus can do with outer dimensions of about 1.50 m in height, 0.5 m in width, and 1.10 m in depth ($1.5 \times 0.5 \times 1.1$ m$^3 = 0.825$ m^3).

The choice of plate size not only depends on the above mentioned criterion [$(A_{envelope}/V)_{min}$], but also on the cross-section of the plate ports. If too small plates had been chosen for a large flowrate, then the velocity in the channels could still be kept reasonably low by a correspondingly larger number of parallel plates; but the velocity in the distributor and collector ducts may, however, become unduly high (danger of erosion). It should be kept below 4 m/s for pure liquids, and less than 2.5 to 3 m/s if solid particles are suspended in the liquid [P2].

In our example, for $S_d = S_c = (\pi/4)D_P^2$ and port diameter of 0.15 m, the inlet velocity on the freshwater side becomes

$$w_{in} = \frac{\dot{V}_2}{(\pi/4)D_P^2} \tag{3.43}$$

$$w_{in} = 1.41 \text{ m/s}$$

With the smallest plates (size 1 from Table 3.1) with $D_P = 0.08$ m, one would get an inlet velocity of nearly 5 m/s! The total pressure drop might be around 1.5 bars on each side for the pack of 105 plates. The pumping powers, therefore, become

$$\dot{W}_{P1} \approx 1.5 \text{ kW} \qquad \dot{W}_{P2} \approx 3.75 \text{ kW}$$

and, with a pump efficiency of 70%, the electrical power required is 7.5 kW.

3.3.2 Shell-and-tube heat exchanger. The problem posed in section 3.1 can be solved with other types of heat exchangers, too. For the sake of comparison, we will perform preliminary calculations for the design of a multipass shell-and-tube heat exchanger in the following steps: Using an estimated overall heat transfer coefficient of $k^{(0)} = 1200$ [W/(m^2 K)], we first obtain from eq. (3.35) a minimum required surface area of

$$A_{req,min}^{(0)} \approx \frac{2\,000}{1\,200} \, 48 \text{ m}^2 = 80 \text{ m}^2$$

With an outer tube diameter of $d_0 = 25$ mm, a total length of tubes $nL = A/(\pi d_0)$ of about 1000 m would be required. In order to reach a flow velocity of around $w_1 \approx 1$ m/s of the wastewater stream inside the tubes (see eq. [3.17]), the number of tubes to be connected in parallel is

$$n_p = \frac{4\dot{V}_1}{w_1 \pi d_i^2} \tag{3.44}$$

With $d_i = 20$ mm, one finds $n_p \approx 32$, i.e., with four tubeside passes, $n = 4n_p = 128$ tubes. The required surface area would, in this case, be obtained with a tube length of $L = 8$ m. To arrive at a more compact design, it is better to choose *eight passes* of 32 tubes each, i.e., 256 tubes with a length of $L = 4$ m. With a relative tube pitch of $s/d_0 = 1.2$, the shell diameter has to be roughly $D_i \approx 550$ mm (problem!).

The flow velocity of the freshwater on the shell-side can be controlled within certain limits by baffles. For pure longitudinal flow without baffles, the shell-side cross-section in the bundle is

$$S_s \approx \frac{\pi}{4}(D_i^2 - nd_o^2) \tag{3.45}$$

$$S_s \approx 0.11 \text{ m}^2$$

leading to a flow velocity of

$$w_{2,\text{longitudinal}} = \frac{\dot{V}_2}{S_s} = 0.223 \text{ m/s}$$

for one pass on the shell-side. Now it has to be checked whether the required efficiency of $\epsilon_1 = 0.833$ can be reached with one shell-side pass and eight tube-side passes: From Table 2.3, for $2m = 8$, $X = 0.4Y$ ($Y = N_1$ and $Z_4 = [(0.4)^2 + (1/4)^2]^{1/2}Y = 0.4717\, Y$, the calculation formula is

$$\frac{1}{\Theta} = \varphi(0.4717\,Y) + \varphi(Y) - \varphi\left(\frac{Y}{4}\right) + \left(0.4 + \frac{1}{4} - 0.4717\right)\frac{Y}{2} \tag{3.46}$$

$$\varepsilon_Y = \varepsilon_1 = Y\Theta$$

$Y = N_1$	2.3	3	4	4.1	5	6	20	∞
$1/\Theta$	2.969	3.694	4.806	4.921	5.976	7.181	24.49	∞
$\varepsilon_Y = \varepsilon_1$	0.775	0.812	0.832	0.833	0.837	0.836	0.817	0.763

The calculations show that $\epsilon_Y = \epsilon_1 = 0.833$ can indeed be achieved with the chosen configuration. However, this requires 4.1 transfer units, instead of the minimum of 2.31 required in pure counterflow. The factor F now is $2.31/4.1 = 0.563$. This means that 77.5% additional surface area would be needed (e.g., tubes of 7m length in place of 4m) or correspondingly higher heat transfer coefficients had to be obtained. Still higher efficiencies could be attained by the use of multiple passes also on the shell-side. This could be realized by a longitudinal shell-side baffle or by connecting two smaller heat exchangers in a countercurrent series. A more accurate recalculation according to [V1] or [H3] can show which actual size of these apparatuses is finally needed (problem!).

3.4 Discussion of Results

The two examples have shown that, in every single case, it has to be checked whether losses in the mean temperature difference ΔT_M can be compensated for by gains in the overall heat transfer coefficients, if heat exchangers with a multipass flow configuration are used. Under the conditions chosen in this example, a plate heat exchanger can usually be more compact and economically more favorable than a shell-and-tube exchanger. The operational limits (pressure, temperature) of the plate heat exchanger are considerably narrower, however, than those of the shell-and-tube type.

4 HEAT EXCHANGERS IN THE FLUE GAS CLEANING PROCESS OF POWER PLANTS EXAMPLE: REGENERATOR

4.1 Description of the Flue Gas Cleaning Process

The intensified requirements of environmental compatibility of our energy supply have led to the need to remove to a high degree, the noxious substances from the flue gas of power plants using oil or coal before they can be led off to the atmosphere through the stack. Today, mainly two process steps are applied: The Flue Gas Desulfurization Plant (FDP) and a plant to reduce the nitric oxides (NO_x), the "DeNO$_x$" plant. Both exist in a number of process variants. Here, as an example, we choose the process described by Klapper, Linde, and Müller [K3] (see Fig. 3.9).

First the SO_2 (along with other noxious substances such as HF, HCl, and heavy metals) is removed from the flue gas in the FDP by scrubbing with water and using a regenerative absorption-desorption circuit. Therefore, the flue gas has to be cooled down first and then heated up again with the incoming raw fluid gas (by the regenerator R1). It leaves the FDP more-or-less dustfree at a temperature of about 80 °C. For the selective catalytic reduction of the nitric oxides (NO_x) still contained in it, with ammonia in a fixed bed reactor, it has to be heated to about 350 °C. The inverse sequence of the two process steps—i.e., first DeNO$_x$, then FDP—would, of course, be more favorable from the viewpoint of thermodynamics as no reheating would be necessary. But the noxious substances in the gas before passing the FDP (dust, HF, HCl, heavy metals, and SO_2) would rapidly poison the catalyst of the DeNO$_x$ reactor or require other, considerably more expensive and less sensitive catalysts.

The flow-sheet contains a total number of six heat exchangers, two of them carried out as regenerators here. Figure 3.10 shows the construction of a DeNO$_x$ plant according to [K3]. Since the unit has often to be retrofitted into an existing power plant, the regenerator and the catalyst beds are arranged one upon the other in a tower construction. The arrangement of two catalytic beds connected in parallel is chosen to keep the pressure drop low. Due to the large volumetric flow rates of the

Purified stack gases

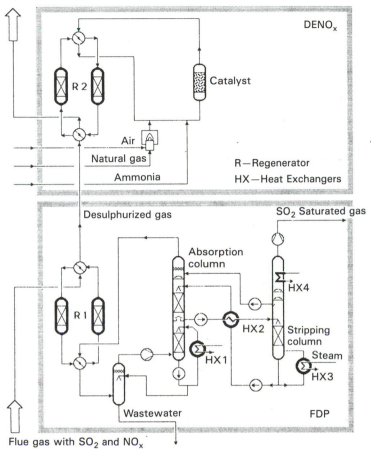

Figure 3.9 Flowsheet of flue gas cleaning process (as described in [K3]).

flue gas, the operational costs for the flue gas blower can be kept within economically reasonably limits only through low pressure drops in all parts of the process. The flue gas streams are in the order of 100 000 to 300 000 m_N^3/h in these units. In larger power plants, the flue gas cleaning may be subdivided into several parallel process paths.

4.2 The Thermal Task

Consider, for example, a flue gas stream of 160,000 m_N^3/h at 80 °C coming from the desulfurization plant to be heated to 330 °C. The temperature is then raised by another 20K to the reaction temperature of 350 °C by a natural gas burner. The hot flue gas, more or less free from NO_x at 350 °C, shall be used to preheat the gas coming from

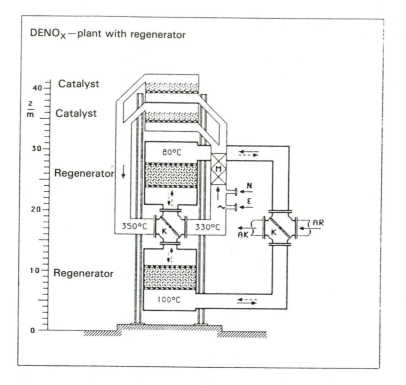

Figure 3.10 Plant for reduction of nitric oxides from flue gases ("DeNOx") with regenerator, design as in Klapper, Linde, Müller [K3]. **M** static mixer, **N** NH_3, **E** natural gas, air, **K** valve, **AR** flue gas from FDP, **AK** flue gas to the stack.

the FDP. Thereby, it shall be cooled to about 100 °C. If we approximate the properties of the mixture of N_2, CO_2, H_2O . . . by those of air, for $\dot{M} \approx 56.7$ kg/s and $c_p \approx 1.03$ kJ/(kg K), we obtain the heat duty as $\dot{Q} = 14.6$ MW.

4.3 Feasibility

Due to the large flue gas streams and the low overall heat transfer coefficients in gas-to-gas heat transfer ($k \approx 5 . . . 35$ [W/(m^2 K)]), very large exchangers will be required. Since the efficiency ϵ has to be very high—it being in our example, $\epsilon = (330 - 80)/(350 - 80) = 0.926$—only counterflow or cross-counterflow configurations come into consideration. Possible types of design are, for example,

- bundles of plain tubes in multiple, counter-directional, cross-counterflow (**gas-to-gas**)
- coupling of two finned-tube bundles by a circulating heat carrier (multiple cross-counterflow) (**gas-to-liquid-to-gas**)

- coupling of two finned-tube bundles by heat pipes (**gas-to-liquid/vapor/liquid-to-gas**)
- rotating regenerator (**gas-to-solid-to-gas**)
- regenerator with fixed masses (**gas-to-solid-to-gas**)

Each of these types has specific advantages and drawbacks, which should be checked and considered in every particular case of design. Very compact exchangers with small flow channels cannot be applied here, because the flue gas still does contain dust, in spite of the pretreatment; and, therefore, the exchangers have to be cleaned from time to time (if possible during operation). Substances like $(NH_4)_2SO_4$ and NH_4HSO_4 may condense on the cooling surfaces as a liquid as viscous as honey.

As pressure drop has to be kept low, this requires low flow velocities and, therefore, large cross-sections. The coupling by a circulating heat carrier may lead to more compact exchangers, but it has the disadvantage that the catalyst might get contaminated in case of a leakage of the heat carrier circuit (oil). An exact tuning of the circulating stream between the flow capacities of the two gas streams is required (see chapter 2, section 6, Fig. 2.44). Hence, measurement and control may also be expensive.

Coupling of two finned-tube bundles by heat pipes, i.e., by internal natural convection streams with evaporation and condensation in closed tubes, is certainly an elegant solution for gas-to-gas heat transfer. In order to reach the required counter-flow effect, the individual rows of heat pipes have to operate at the corresponding local mean temperatures between the two streams, i.e., at the cold end at 90 °C and at the hot end at 340 °C in our example. With water as a working fluid, the inside pressure of the heat pipes at the hot end would have to be about 150 bars.

Regenerators with fixed storage mass have the advantage of a relatively simple and robust design (e.g., simple packed columns). A particular disadvantage is the switching losses: At each switching, at least one gas volume (regenerator hold-up) passes the plant uncleaned. This limits the possible degree of removal of noxious components but may be kept very small, however, in practice. These switching or rinsing losses also occur in rotating regenerators. In addition, the leakage past the sliding gaskets compounds the problem of loss of fluid.

As an exercise it is recommended that the design calculations be redone for all these types of heat exchangers. As an example, we show the design of a regenerator with fixed masses in the following (see Fig. 3.10).

4.4 Design of a Regenerator with Fixed Storage Masses

According to eq. (2.172), the ideal regenerator (without longitudinal conduction and with $N_s = 0$) requires the number of transfer units

$$N_{req,min} = \frac{2\varepsilon}{1 - \varepsilon} \qquad (3.47)$$

With the specified temperatures (see Fig. 3.10), one obtains $N_{req,min} = 25$ and, thereby,

$$A_{\text{req,min}} = N_{\text{req,min}} \frac{\dot{M} c_{\text{p}}}{k} \tag{3.48}$$

$$A_{\text{req,min}} = 1.46 \cdot 10^6 \text{ m}^2 \text{ (W/m}^2 \text{ K)}/k.$$

The heat transfer coefficient k (gas-to-solid) is here in the range of 10 . . . 100 [W/(m^{2}K)], so that transfer surface areas in the order of 15 000 to 150 000 m^2 will be required. If we choose a bed of 50mm ceramic saddles (see Fig. 3.11) with a voidage of $\psi = 73\%$ and a volume specific surface of $a_{\text{v}} = 120$ m^2/m^3, we obtain from this a bed volume of 125 . . . 1 250 m^3 (a cube with edge length of 5 . . . 11 m).

To keep pressure drop low, a sufficiently large flow cross-section is chosen, e.g., $S = (6 \times 9) \text{ m}^2 = 54 \text{ m}^2$. Then, the velocity in the bed

Packing	Size	Bed density ρ_{bed}	Volume number n_{v}	Specific surface a_{v}	Void-fraction ψ
	mm	kg/m^3	(dm)$^{-3}$	m^{-1}	%
Raschig-Ring	15×15×2	700	210	330	70
	25×25×3	620	46	195	73
	50×50×5	520	6,4	98	78
	100×100×10	450	0,75	44	81
Pall-Ring	25×25×3	620	46	220	73
	50×50×5	550	6,3	120	78
Intalox-Saddle	15	670	400	450	71
	25	610	85	255	74
	50	530	9,3	120	79
Berl-Saddle	15	800	280	430	67
	25	700	75	260	69
	50	600	8	120	73

Figure 3.11 Characteristic data of ceramic packing material.

$$w_\psi = \frac{\dot{M}}{\varrho \psi S} \qquad (3.49)$$

with $\rho_{min} = \rho(350\,°C) \approx 0.56$ kg/m^3 becomes less than about $w_{\psi,max} \approx 2.6$ m/s.

Calculation of the gas-side heat transfer coefficient

The gas-side heat transfer coefficient is calculated, according to Gnielinski [G8; V1, Gh1; H3]:

$$\boxed{Nu_{bed} = f_a\, Nu_{single\ sphere}} \qquad (3.50)$$

$f_a = 1 + 1.5(1 - \psi)$ for spheres,
$f_a = 1.6$ for finite cylinders and cubes,
$f_a = 2.1$ for hollow cylinders (Raschig-rings),
$f_a = 2.3$ for saddles (Berl-saddles).

For a single sphere, one has

$$Nu_{single\ sphere} = 2 + (Nu_{lam}^2 + Nu_{turb}^2)^{1/2} \qquad (3.51)$$

with

$$Nu_{lam} = 0.664 Pr^{1/3}\, Re^{1/2}$$

$$Nu_{turb} = \frac{0.037 Re^{0.8}\, Pr}{1 + 2.44 Re^{-0.1}(Pr^{2/3} - 1)}$$

The characteristic length in the dimensionless numbers Nu and Re is the diameter d_S of a sphere with the same surface area:

$$d_S = \left(\frac{A_P}{\pi}\right)^{1/2} \qquad (3.52)$$

The characteristic velocity in Re is w_ψ from eq. (3.49). The surface of a particle A_P may be easily calculated from the data given by the manufacturer. The volume specific surface area a_V and the volume specific number n_V:

$$A_P = \frac{a_V}{n_V} \qquad (3.53)$$

With the data from Fig. 3.11, one obtains for the 50 mm Berl-saddles:

$$d_S = 69\ mm$$

The physical properties of flue gas ($\approx$ as for air) at the mean temperature of $\approx 220\,°C$ are [V1]

$$\rho \approx 0.71 \text{ kg/m}^3$$

$$\left.\begin{array}{l} c_p \approx 1.03 \text{ kJ/kgK} \\ \lambda \approx 0.0403 \text{ W/Km} \\ \eta \approx 26.4 \cdot 10^{-6} \text{ Pas} \end{array}\right\} \quad Pr \approx 0.67$$

From this, with $\dot{M} = 56.7$ kg/s,

$Re = 3759$ $\qquad$ $Nu_{\text{lam}} = 35.62$ $\qquad\qquad$ $Nu_{\text{turb}} = 23.98$

$\qquad\qquad\qquad$ $Nu_{\text{single sphere}} = 44.94$ $\qquad\qquad$ $Nu_{\text{bed}} = 103.4$

$\qquad\qquad\qquad$ $\alpha_g = (\lambda/d_s)Nu_{\text{bed}}$

$$\boxed{\alpha_g = 60.4 \text{ W/m}^2\text{K}}$$

Estimation of the internal heat conduction resistance

The solid volume of one saddle particle is

$$V_P = \frac{1 - \psi}{n_V} \tag{3.54}$$

and, with eq. (3.53), the ratio of volume to surface becomes

$$\frac{V_P}{A_P} = \frac{1 - \psi}{a_V} \tag{3.55}$$

i.e., $V_P/A_P = [(1 - 0.73)/120]$ m $= 2.25$ mm. The mean wall thickness s_P of the saddle is, therefore, about 4.5 mm. The internal heat transfer coefficient is greater than [H3, V1]

$$\alpha_i \geq 6 \frac{\lambda_P}{s_P} \tag{3.56}$$

With an estimated heat conductivity of the ceramic particles of $\lambda_P = 1$ W/(Km), we get from this

$$\boxed{\alpha_i \geq 1\,333 \text{ W/m}^2\text{K}}$$

The overall heat transfer coefficient k then becomes

$$\frac{1}{k} = \frac{1}{\alpha_g} + \frac{1}{\alpha_i} \tag{3.57}$$

about $k = 57.8$ [W/(m²K)]. The required transfer surface area from eq. (3.48) results in $A_{\text{req}} = 25\,260$ m² and the bed volume $V = A/a_V = 210$ m³. With the chosen base area of $S = 54$ m², we, therefore, find a bed height of $L = V/S \approx 3.90$ m.

Determination of the period (switching time of the regenerator)

On one hand, the period has to be short enough, so that the deviation from an ideal regenerator remains within reasonable limits; on the other hand, it has to be chosen long enough to minimize switching losses. The latter condition means that the duration of the period must be much longer than the residence time of the gas in a bed:

$$t_{1,2} \gg t_{R,g} = \frac{L}{w_\psi} \tag{3.58}$$

The gas residence time is roughly two seconds. The first condition means that the number of transfer units of the storage mass N_S has to be kept small:

$$t_{1,2} = \frac{(\varrho c)_P V_P}{k A_P} N_S \tag{3.59}$$

The time constant in N_S is thereby ($N_S = t_{1,2}/t_{R,S}$).

$$t_{R,S} = \frac{(1 - \psi)(\varrho c)_P}{k a_V} \tag{3.60}$$

With $\psi = 0.73$, $k = 57.8$ [W/(m² K)], $a_V = 120$ m⁻¹, and $(\rho c)_{P,\text{ceramics}} \approx 2 \cdot 10^6$ J/(m³ K), one gets

$$t_{R,S} = 77.9 \ s$$

If a duration of the period of $t_{1,2} = 6$ min is chosen, as given in [K3], then

$$N_S = \frac{6 \cdot 60}{77.9} = 4.62$$

From this, with $N = 25$ from eq. (2.175), one gets the LMTD-correction factor F:

$$F \approx 1 - \frac{(4N_S/5) - 3 \tanh(N_S/5)}{N} = 0.939$$

This deviation from the ideal regenerator can be accounted for by slightly increasing the bed height ($NF = N_{\text{req,min}}$):

$$N \approx N_{\text{req,min}} + [(4N_S/5) - 3 \tanh (N_S/5)] \tag{3.61}$$

$$N \approx 25 + 1.51 = 26.51$$

In practice, one will choose a bed height of about 4.10 m in place of the 3.90 m calculated for the ideal regenerator. The whole arrangement of two packed beds with the valves and the flue gas ducts will have roughly the dimensions shown in Fig. 3.10.

Calculation of pressure drop

Pressure drop can be calculated approximately from Ergun's equation [B5, H3, V1]

$$\Delta p = \xi \frac{\varrho}{2} w^2 \frac{L}{d_P} \tag{3.62}$$

$$\xi = \frac{1 - \psi}{\psi^3} \left(\frac{300(1 - \psi)}{Re} + 3.5 \right)$$

$$d_P = 6 \frac{V_P}{A_P} = \frac{6(1 - \psi)}{a_V}$$

$$Re = \frac{\varrho w d_P}{\eta}$$

$$\varrho w = \frac{\dot{M}}{S}$$

From this, one obtains

$$w = 1.479 \text{ m/s,}$$
$$d_P = \frac{6(1 - 0.73)}{120} \text{ m} = 13.5 \text{ mm}$$
$$Re = 536.9 \qquad \xi = 2.534 \qquad \Delta p = 597.6 \text{ Pa.}$$

For two beds, therefore, $\Delta p \approx 1\,200$ Pa. With the volume flowrate of $\dot{M}/\rho \approx 80$ m³/s, this leads to a pumping power of $\dot{W}_p \approx 96$ kW to overcome the flow resistances of just the packed beds.

4.5 Discussion of Results

In a more accurate recalculation, one should account for the fact that, due to the additional heating with the natural gas burner, the streams are no longer exactly equal. The hot stream is slightly stronger, i.e., it will correspondingly not cool down to 100 °C, but only to a somewhat higher temperature. Also, the temperature dependency of the physical properties might be taken into account in a more detailed recalculation. The calculation could also be repeated with other data (bed material, bed cross section) and for other types of heat exchangers for comparison, in order to find the most favorable solution for a given problem in each particular case. The heat exchangers HX 1–4 in Fig. 3.9 are mainly designed as plate heat exchangers [K3]. For their design, the flow rates of water, wash-solution, cooling- and hot-water streams have to be known. For the design of absorption and desorption columns, see, e.g., [S3].

5 EVAPORATION COOLING
EXAMPLE: FALLING FILM CONDENSER
FOR REFRIGERANTS

5.1 Problem Statement

In the condenser of refrigerant cycles (refrigerators, heat pumps, heat transformers [H1, S5]), a heat rate $\dot{Q}_C$ has to be transferred to the surroundings corresponding to the product of the mass flow rate of refrigerant to be condensed, and the specific enthalpy of vaporization.

$$\dot{Q}_C = \dot{M}_C \Delta h_{v,C} \tag{3.63}$$

Eq. (3.63) is an energy balance for the condensate side, assuming that a pure refrigerant enters the condenser as a saturated vapor and is withdrawn from the apparatus as a condensate without significant subcooling. Then the temperature of the condensing refrigerant remains constant, as long as the slight change in condensing temperature due to the pressure drop on the condensate side may be neglected.

As an example, we will design a condenser for a heat duty of $\dot{Q}_C = 580$ [kW] for ammonia at a condensing temperature $T_c = 40\,°C$, corresponding to a saturation pressure of NH_3 of about 16 bars. As a coolant, water is available at a maximum volume flowrate of 10 m³/h and a temperature of $T' = 15\,°C$. The apparatus is to be designed for operation in a hot, dry climate (desert climate) with an air temperature of $T_\infty = 45\,°C$ and an absolute humidity of $Y_\infty = 6\cdot10^{-3}$ (Y = mass of water vapor/ mass of dry air).

5.2 Check for Feasibility

Since the condensation temperature is below the temperature of the surrounding air, direct air cooling is impossible. The cooling water available could take up a maximum heat load of

$$\dot{Q}_{W,max} = \dot{M}_W c_{pW}(T_c - T') \tag{3.64}$$

if it is heated up to $T'' = T_c = 40\,°C$ in the limit, i.e., with $c_{pW} = 4.19$ [kJ/(kg K)], $\rho_W = 1000$ kg/m³,

$$\dot{Q}_{W,max} = 10 \text{ m}^3 \cdot 1\,000 \text{ kg/m}^3 \cdot 4.\,19 \text{ kJ/kgK } (40-15) \text{ K}/3600 \text{ s}$$

$$\dot{Q}_{W,max} = 291 \text{ kW}$$

Thus, water cooling will not be adequate for the required condensation duty. We will check if evaporative cooling is feasible here. If the water temperature would rise to, say, $T_W = 35\,°C$, for which the saturation moisture content is $Y^*(35\,°C) = 37\cdot10^{-3}$, there is a considerable driving force for evaporation.

5.3 Energy Balances, Rate Equations

Figure 3.12 shows schematically the principle of a falling film condenser with the cooling water streams $\dot{M}_F$ = feed stream, $\dot{M}_R$ = recirculated stream, $\dot{M}_V$ = evaporation stream (vapor). At steady state, the cooling water side energy balance is

$$\dot{Q}_C + \dot{M}_F c_{pw} T' + \dot{Q}_A - \dot{M}_V h_V(T_{film}) - (\dot{M}_F - \dot{M}_V) c_{pw} T'' = 0 \qquad (3.65)$$

Here we have made use of the mass balance $\dot{M}_{out} = \dot{M}_F - \dot{M}_V$. A balance around the mixing junction of feed and recirculation streams gives

$$\dot{M}_F c_{pw} T' + \dot{M}_R c_{pw} T'' = (\dot{M}_F + \dot{M}_R) c_{pw} T_M \qquad (3.66)$$

As an approximate mean film temperature between T_M (top) and T'' (bottom), we use the arithmetic mean

$$T_{film} = \frac{T_M + T''}{2} \qquad (3.67)$$

Eq. (3.65) can also be written as

$$\dot{Q}_C = \dot{M}_F c_{pw}(T'' - T') - \dot{Q}_A + \dot{M}_V \Delta h \qquad (3.65a)$$

The rate equations, or heat transfer kinetics, are

$$\dot{Q}_C = k_{CW} A(T_c - T_{film}) \qquad (3.68)$$

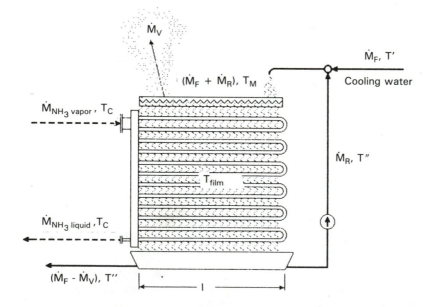

Figure 3.12 Falling film condenser.

and

$$\dot{Q}_A = k_{AW}A(T_\infty - T_{film}) \qquad (3.69)$$

The evaporation rate is formulated with a corresponding equation for mass transfer kinetics [S2]:

$$\dot{M}_V = \rho_A \beta_A A[Y^*(T_{Ph}) - Y_\infty] \qquad (3.70)$$

The conductive resistance in the falling film can be neglected; and the temperature at the film surface (phase interface) T_{Ph} can, therefore, be put equal to the mean film temperature T_{film}, with a reasonable degree of approximation. The overall heat transfer coefficient k_{AW} from the air to the water film, thus, becomes practically equal to the outer air side heat transfer coefficient α_A. Regarding $\dot{Q}_C$, k_{CW}, α_A, β_A, T_c, T', $\dot{M}_F$, c_{pW}, η_W, ρ_W, T_∞, Y_∞, ρ_A, $Y^*(T)$, and $\Delta h(T)$ as given quantities, with eqs. (3.65)–(3.70), one has six equations to determine the unknowns A, $\dot{M}_V$, $\dot{M}_R$, $\dot{Q}_A$, T_{film}, T'', and T_M. Since these are seven unknown quantities, it is necessary to choose one additional relation between the unknowns and the given quantities.

5.4 Determination of the Film Mass Velocity

The recirculation rate $\dot{M}_R$ should be chosen such that the tubes are wetted uniformly by a thin but continuous film of water. If the film mass velocity $\dot{B} = (\dot{M}_F + \dot{M}_R)/l$ (mass flowrate per length l of the horizontal tubes) is too small, the film will break up into single rivulets, and the surface is not completely wetted; if $\dot{B}$ is too large, the liquid will splash away from the tubes, which would be useless for the cooling process. If the film mass velocity is divided by the liquid density and by twice the mean film thickness $2 \cdot s_W$ (the film flows along the perimeter of a tube on **both** sides), a mean flow velocity of the falling film results:

$$w_{film} = \frac{\dot{B}}{2s_W \cdot \rho_W} \qquad (3.71)$$

With this, we may define a film Reynolds number in the form

$$Re_f = \rho_W w_{film} \frac{s_W}{\eta_W} = \frac{\dot{M}_R + \dot{M}_F}{2l\eta_W} \qquad (3.72)$$

A favorable film mass velocity—uniform wetting, no splashing—is found from experience at a film Reynolds number of

$$Re_f \approx 300 \qquad (3.73)$$

The film mass velocity should, therefore, be chosen to be about

$$\dot{B} \approx 600\eta_W \qquad (3.74)$$

i.e., for **water** with $\eta_W \approx 10^{-3}$ [Pas]

$$\dot{B} \approx 0.6 \text{ kg/(ms)} \approx 2\,000 \text{ kg/(mh)}$$

5.5 Choice of Dimensions of Tubular Element

A row of tubes as in Fig. 3.12 may be built from 48 parallel tubes (24 U-tubes) of yet unspecified length l with $d_o \times s = 38 \times 3.5$ [mm × mm]. With a tube pitch of $2d_o$, this results in a heat exchanger element of height H $= 48 \times 2 \times 38$ [mm] $= 3.648$ [m]. The transfer surface area A per length l of the element, thus, becomes

$$\frac{A}{l} = 48\pi d_o = 5.730 \text{ m}^2/\text{m}$$

Together with eq. (3.74), one can fix the film flux (mass flowrate divided by the surface area) as a constant

$$\dot{m}_B = \frac{(\dot{M}_F + \dot{M}_R)/l}{A/l} = \text{const} \tag{3.75}$$

From the numerical values, we get

$$\dot{m}_B = \frac{2\,000 \text{ kg/(hm)}}{3\,600 \text{ (s/h) } 5.73 \text{ m}^2/\text{m}}$$

$$\dot{m}_B = 0.09696 \text{ kg/(m}^2\text{s)}$$

5.6 Calculation of Reflux and Surface Area Required

The key to the determination of surface area and reflux required is the mean film temperature T_{film} (see eqs. [3.65a], [3.67], [3.70]). For abbreviation and for elimination of unknowns, it is convenient to introduce a recirculation ratio v:

$$v \equiv \frac{\dot{M}_F + \dot{M}_R}{\dot{M}_F} = \dot{m}_B \frac{A}{\dot{M}_F} \tag{3.76}$$

With it, the water outlet temperature T'' and the mixing temperature T_M can be expressed in terms of the T_{film} and the feed water inlet temperature T' (eqs. [3.66] and [3.67]):

$$T'' - T' = \frac{2v}{2v - 1}(T_{film} - T') \tag{3.77}$$

With this prelude, the balance eq. [3.65a] can be written as an equation for T_{film} (after division by A and with eqs. [3.68]–[3.70], [3.75]–[3.77]):

$$k_{CW}(T_c - T_{film}) = \frac{2\dot{m}_B c_{pW}}{(2v - 1)}(T_{film} - T') - \alpha_A(T_\infty - T_{film})$$
$$+ \rho_A \beta_A [Y^*(T_{film}) - Y_\infty]\Delta h_V$$

(3.78)

or shorter

$$\dot{q}_C = \dot{m}_F \Delta h_W - \dot{q}_A + \dot{m}_V \Delta h_V \qquad (3.78a)$$

The transfer surface area A is related simultaneously to the recirculation ratio v (from eq. [3.76]), and with T_{film} via eq. [3.68],

$$A = \frac{\dot{M}_F v}{\dot{m}_B} \qquad (3.76a)$$

$$A = \frac{\dot{Q}_C}{k_{CW}(T_c - T_{film})} \qquad (3.68a)$$

From these two equations, a second relationship between v and T_{film} is obtained

$$v = \frac{\dot{Q}_C \dot{m}_B}{\dot{M}_F k_{CW}(T_c - T_{film})} \qquad (3.79)$$

From eqs. (3.78) and (3.79), the two quantities v and T_{film} can now be calculated.

Estimation of the transfer coefficients

The overall heat transfer coefficient from the condensing ammonia to the falling water film is estimated to be $k_{CW} \approx 600$ [W/(m² K)] [V1]. A recalculation can be done later with the methods given in the handbooks [V1, H3]. The air side heat transfer coefficient α_A and the mass transfer coefficient β_A, related to it via Lewis' law, are chosen here as

$$\alpha_A \approx 30 \text{ [W/(m}^2 \text{ K)]}$$

and

$$\rho_A \beta_A \approx \alpha_A/(c_{pA} + Y_\infty c_{pV})$$

The numerical values obtained for the individual terms of eqs. (3.78) and (3.79) are given in the table below. From these are found, iteratively, the mean film temperature and recirculation ratio and, thus, the surface area A and so the length l, i.e., the whole dimensions of the apparatus (see Fig. 3.13)

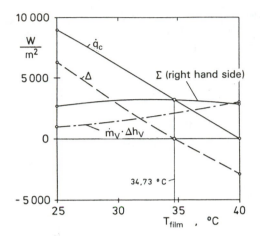

Figure 3.13 Graphic determination of the film temperature.

$T_{film}/[°C]$	25	30	34.7	34.8	35	40
$Y^*/[g/kg]$	20.3	27.6	36.4	36.6	37.1	49.6
v	2.25	3.37	6.37	6.49	6.75	∞
$\dot{m}_F \Delta h_W/[W/m^2]$	2321	2119	1364	1342	1300	0
$\dot{m}_V \Delta h_V/[W/m^2]$	1000	1497	2103	2118	2148	2997
$-\dot{q}_A/[W/m^2]$	−600	−450	−309	−306	−300	−150
Σ (right-hand side)	2721	3166	3158	3154	3147	2847
$\dot{q}_c/[W/m^2]$	9000	6000	3180	3120	3000	0
$\Delta/[W/m^2]$	6279	2834	22.1	−34.5	−147	−2847

The difference Δ between the values of the left and the right sides of eq. (3.78) from this becomes zero at a mean film temperature of

$$T_{film} = 34.73\,°C$$

From this follows

$$\dot{q}_c = 3\,162\ W/m^2$$
$$A = 183.4\ m^2$$

and

$$l = 32.0\ m$$

i.e., eight tube row elements with a length of 4 m each. The recirculation rate is $v = 6.40$, i.e., the reflux stream $\dot{M}_R$ must be chosen to be 5.4 times the feed stream $\dot{M}_F$:

$$\dot{M}_R = 5.4(1\,000 \text{ kg/m}^3)10 \text{ m}^3/(3\,600 \text{ s}) = 15 \text{ kg/s}$$

The calculation should be repeated with calculated values of the transfer coefficients.

6 EVAPORATION AND CONDENSATION EXAMPLE: HEAT EXCHANGER WITH VERTICAL TUBES, STEAM HEATED INTERNALLY [M3]

6.1 Problem Statement

Vertical tubes, internally heated by condensing steam ("heater candles") are used in many applications to heat up or to evaporate liquids. Figure 3.14 shows a bayonet heat exchanger with vertical tubes (Field tubes) offered as standard items by some manufacturers. In these apparatuses, steam flows upward through the inner tubes; condenses at the inner wall of the outer tubes, which are closed at the top; and forms a condensate film flowing downward by gravity. The enthalpy ("latent heat") released as heat in condensation flows past the wall of the tubes to the liquid in the heated vessel and vaporizes it, if sufficiently high temperatures are reached.

The design of such an apparatus, or its sizing, i.e., determination of the transfer surface area required for given heat duty, at first sight seems to be rather simple, because the driving temperature difference $\Delta T = T_C - T_V$ is constant over the whole surface area (so long as the pressure drop of steam flow as well as the static pressure difference of the liquid over the height are neglected). Once the mean overall heat transfer coefficient k is known, the surface area required is easily calculated from

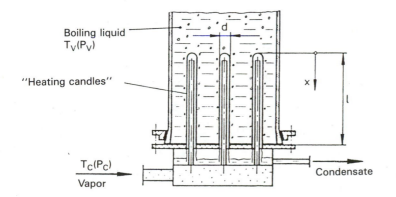

Figure 3.14 Bayonet heat exchanger as a reboiler.

$$A = \frac{\dot{Q}}{k \triangle T} \qquad (3.80)$$

in which $\dot{Q}$ is the heat duty to be transferred. The difficulties here are only in the determination of this mean overall heat transfer coefficient k. First, the local overall heat transfer coefficient k_{loc} has to be calculated from the individual heat transfer coefficients—α_C on the condensate side, α_w for conduction through the wall, and α_v on the vaporization side.

$$\boxed{\frac{1}{k_{loc}} = \frac{d_o/d_i}{\alpha_{C\ loc}} + \frac{d_o/d_m}{\alpha_w} + \frac{1}{\alpha_{V\ loc}}} \qquad (3.81)$$

$$k_{loc} = \frac{d\dot{Q}}{\triangle T dA_o} = \frac{\dot{q}_{loc}}{\triangle T}$$

The heat transfer coefficients α_c and α_v, in turn, depend on the local heat flux $\dot{q}_{loc} = k_{loc}\triangle T$. For the vaporization side, one can write

$$\boxed{\alpha_{V\ loc} = c\dot{q}_{loc}^n} \qquad (3.82)$$

with $n = 0.6 \ldots 0.8$ for pool boiling, $n = 0.25$ for natural convection, and $n = 0$ for forced convection. According to Nusselt [N1], one finds for laminar film condensation

$$\boxed{\alpha_{C\ loc} = \frac{\lambda_f}{s(x)}} \qquad (3.83)$$

where the local thickness of the condensate film $s(x)$ is related to the local heat flux via the differential equation [N1]

$$\boxed{\frac{d\dot{Q}}{dA_i} = \dot{q}_{loc}\frac{d_o}{d_i} = \frac{g\varrho_f \triangle h_v}{v_f}s^2\frac{ds}{dx}} \qquad (3.84)$$

To calculate only the local overall heat transfer coefficient k_{loc}, a simultaneous solution of eqs. (3.81)–(3.84) is required. To come to the mean overall heat transfer coefficient k eventually, another integration has to be performed:

$$k = \frac{1}{l}\int k_{loc}(x)\,dx \qquad (3.85)$$

In the following, we will first show how to calculate the mean overall heat transfer coefficient k in a commonly used approximation from mean heat transfer coefficients α_C, α_V; subsequently, we shall derive and discuss a rigorous solution of eqs. (3.81) to (3.84).

6.2 Approximate Calculation of Mean Overall Heat Transfer Coefficients

Writing an approximate relation for the *mean* overall heat transfer coefficient k in the form of eq. (3.81), we get

$$\frac{1}{k} \approx \frac{d_o/d_i}{\alpha_C} + \frac{d_o/d_m}{\alpha_W} + \frac{1}{\alpha_V} \qquad (3.81a)$$

In this, α_V is directly expressed as a function of k via $c\dot{q}^n$; and $\dot{q} = k\Delta T$, α_C, however, has to be calculated from eqs. (3.83) and (3.84). Integration of eq. (3.84) yields

$$\frac{s^3}{3} = \left(v_f \dot{q} \frac{d_o/d_i}{g\varrho_f \Delta h_V} \right) x \qquad (3.86)$$

and thus

$$\alpha_{C,loc} = \frac{\lambda_f}{(v_f^2/g)^{1/3}} \left(\frac{\Delta h_V \eta_f}{3\dot{q}(d_o/d_i)x} \right)^{1/3} \qquad (3.87)$$

The term $\dot{q}(d_o/d_i)x/(\Delta h_V \eta_f)$ is the film Reynolds number usually defined in condensation as $Re_{fx} = \dot{M}_{Cx}/(b\eta_f)$, $b = \pi d_i$. Using it and the definition $Nu_{Cx} = \alpha_{C\ loc}(v_f^2/g)^{1/3}/\lambda_f$, eq. (3.87) can be shortened as

$$Nu_{Cx} = (3Re_{fx})^{-1/3} \qquad (3.87a)$$

The mean heat transfer coefficient or the corresponding Nusselt number Nu_C can be obtained from this via

$$Nu_C = \frac{Re_f}{\int dRe_{fx}/Nu_{Cx}(Re_{fx})} \qquad (3.88)$$

to be

$$Nu_C = \frac{4}{3}(3Re_f)^{-1/3} \qquad (3.89)$$

In the original form, this reads

$$\alpha_C = \frac{\lambda_f}{(v_f^2/g)^{1/3}} \frac{4}{3} \left(\frac{\Delta h_V \eta_f}{3\dot{q}(d_o/d_i)l} \right)^{1/3} \qquad (3.89a)$$

With eq. (3.89a) and $\alpha_V = c\dot{q}^n$, eq. (3.81a) yields an equation for the mean heat flux $\dot{q}(l)$:

$$\frac{1}{\dot{q}} \approx \frac{d_o/d_i}{\alpha_C(\dot{q})\Delta T} + \frac{d_o/d_m}{\alpha_W \Delta T} + \frac{1}{c\dot{q}^n \Delta T} \qquad (3.81b)$$

This cannot be solved explicitly for $\dot{q}$. The solution may be found, however, in the form of the inverse function $l(\dot{q})$. With the abbreviations,

$$\dot{q}_{Vm} \equiv (c\triangle T)^{\frac{1}{1-n}}$$

(3.90)

$$l^* \equiv \frac{(g\varrho_f \triangle h_v \lambda_f^3 / v_f)(\triangle T)^3}{(\dot{q}_{Vm} d_o/d_i)^4}$$

(3.91)

$$Z \equiv \frac{\dot{q}_{Vm}}{\dot{q}} = \frac{\alpha_{Vm}}{k} = \frac{c^{\frac{1}{1-n}} \triangle T^{\frac{n}{1-n}}}{k}$$

(3.92)

$$B \equiv \frac{\dot{q}_{Vm} d_o/d_m}{\alpha_w \triangle T} = \frac{\alpha_{Vm} d_o \ln(d_o/d_i)}{2\lambda_w}$$

(3.93)

$$\xi_l \equiv \frac{l}{l^*}$$

one eventually obtains

$$Z \approx \frac{3}{4}\left(\frac{3\xi_l}{Z}\right)^{1/3} + B + Z^n$$

(3.81c)

or, solved for the length,

$$\xi_l \approx \frac{Z}{3}\left[\frac{4}{3}(Z - Z^n - B)\right]^3$$

(3.94)

The result of this approximate solution will be compared with the rigorous solution after the next subsection.

6.3 Calculation of the Mean Overall Heat Transfer Coefficient from the Local Variation

The local overall heat transfer coefficient k_{loc} from eq. (3.81) depends on the local film thickness $s(x)$. The dimensionless condensate thickness is

$$\sigma \equiv \frac{\alpha_{Vm}(d_o/d_i)}{\lambda_f} s(x)$$

(3.95)

The local dimensionless overall heat transfer resistance written in similar form to eq. (3.92) is

$$z \equiv \frac{\dot{q}_{Vm}}{\dot{q}_{loc}} = \frac{\alpha_{Vm}}{k_{loc}}$$ (3.96)

$$\alpha_{Vm} \equiv c^{\frac{1}{1-n}} \Delta T^{\frac{n}{1-n}}$$

With these and the relative conductive resistance of the wall already defined in eq. (3.93), eqs. (3.81) to (3.83) now condense to

$$z = \sigma + B + z^n$$ (3.81d)

and eq. (3.84) with $\xi = x/l*$ becomes

$$\frac{1}{z} = \sigma^2 \frac{d\sigma}{d\xi}$$ (3.84a)

Now eq. (3.84a) could be integrated if $z(\sigma)$ could be explicitly obtained from eq. (3.81d). This, however, can only be solved inversely for $\sigma(z)$:

$$\sigma(z) = z - z^n - B$$ (3.97)

The integration of eq. (3.84a) is, therefore, executed over dz in place of $d\sigma$:

$$d\sigma = \frac{d\sigma}{dz} dz$$

$$d\xi = z\sigma^2 \frac{d\sigma}{dz} dz = f(z) dz$$ (3.98)

$$f(z) = (z - nz^n)(z - z^n - B)^2$$ (3.99)

$$\xi = F(z) - F(z_0)$$ (3.100)

$$F(z) = \int f(z) dz.$$

The integral function $F(z)$ is found by multiplying the terms on the right side of eq. (3.99) and subsequent integration term by term with $\Theta = z^{n-1} = \Delta T_v/\Delta T$:

$$F(z) = z^4 \left(\frac{1}{4} - \frac{2+n}{3+n}\Theta + \frac{1+2n}{2+2n}\Theta^2 - \frac{n}{1-3n}\Theta^3 \right)$$
$$- z^3 2B \left(\frac{1}{3} - \frac{1+n}{2+n}\Theta + \frac{n}{1+2n}\Theta^2 \right) + z^2 B^2 \left(\frac{1}{2} - \frac{n}{1+n}\Theta \right)$$ (3.101)

The mean overall heat transfer coefficient k is then obtained from eq. (3.85)

$$k = \frac{1}{l} \int k_{loc}(x)\,dx \tag{3.85}$$

or, in dimensionless form,

$$\frac{1}{Z} = \frac{1}{\xi_l} \int \frac{1}{z}\,d\xi \tag{3.85a}$$

In this integral, $d\xi/z$ can be substituted from eq. (3.84a) by $\sigma^2 d\sigma$:

$$\frac{1}{Z} = \frac{1}{3\xi_l}(\sigma^3 - \sigma_0^3) \tag{3.102}$$

or, with eq. (3.97) and $\sigma_0 = \sigma(\xi = 0) = 0$:

$$\boxed{\frac{k}{\alpha_{VM}} = \frac{1}{3\xi_l}[z(\xi_l) - z^n(\xi_l) - B]^3} \tag{3.103}$$

(see eq. [3.94] for comparison).

Now, with eqs. (3.100) and (3.103), one has a parametric representation of the function $k(l)$. Equation (3.100) yields $l(z)$ and eq. (3.103) corresponding values of $k(z)$. The relative local overall heat transfer resistance $z = (\alpha_{Vm}/k_{loc}) = \dot{q}_{Vm}/\dot{q}_{loc}$ or, alternatively, the relative local vaporization side temperature difference $\Theta = \Delta T_V/\Delta T = z^{n-1}$ may be used as parameters.

In order to evaluate eq. (3.100), i.e., the relation between the position x and the local overall heat transfer coefficient k_{loc} (or the local heat flux $\dot{q}_{loc}$, or the local vaporization side temperature difference ΔT_V), it is necessary first to calculate $z_0 = z(x = 0)$ from eq. (3.97). This can be done explicitly only for $\sigma_0 = 0$ and $B = 0$, i.e., for vanishing condensation film thickness and vanishing conduction resistance of the tube wall. In this limiting case, one simply obtains $z_0 = 1$ ($k_{loc}(x = 0) = \alpha_{Vm}$). For finite values of σ_0 and/or B, z_0 is best obtained iteratively as the root (zero) of a function $\varphi(z)$,

$$\varphi(z) = z - z^n - B - \sigma_0 \tag{3.104}$$

with

$$\varphi'(z) = 1 - nz^{n-1}$$

by Newton's method:

$$z_{0,\nu+1} = z_{0,\nu} - \frac{\varphi(z_{0,\nu})}{\varphi'(z_{0,\nu})} \tag{3.105}$$

As a starting value, one can simply choose $z_{0,0} = 1$. The iteration converges very rapidly. To calculate $z_1 = z(\xi_1)$, i.e., the local heat flux $\dot{q}_{loc}(l)$ at a certain specified total length l or at a certain position x_l, eq. (3.100) again has to be solved iteratively for z_l:

With

$$\Phi(z) = F(z) - F(z_0) - \xi_l \qquad (3.106)$$

$$\Phi'(z) = f(z) \qquad (3.107)$$

from eq. (3.99) we get

$$z_{l,v+1} = z_{l,v} - \frac{\Phi(z_{l,v})}{\Phi'(z_{l,v})} \qquad (3.108)$$

To calculate the variation of the local and the integral mean heat transfer coefficients and overall heat transfer coefficients between $x = 0$ and $x = l$, the iterations from eq. (3.108) need not be carried out at each positon. One can simply give arbitrary values $z_1, z_2, z_3, \ldots z_k$ (with $z_0 \le z_k \le z_l$) as parameters and calculate the corresponding values of x and k from eqs. (3.100) and (3.103), respectively. The evaluation can be made easier for a user by a graphical presentation of the function $(1/Z) = k/\alpha_{Vm}$ vs. ξ_l with B and n as parameters.

Figure 3.15 shows such plots for (a): $n = 0$ (forced convection); (b): $n = 0.25$ (natural convection); and (c): $n = 0.7$ (pool boiling) in the range $0 \le \xi \le 3$ with relative conduction resistances of the tube wall of $B = 0, 0.25, 0.5$, and 1.0.

Figure 3.16 shows, additionally, for $n = 0.7$ (pool boiling) and $B = 0.25$ the variation of inner and outer wall temperatures and, therefore, the local variation of the individual temperature differences ΔT_V, ΔT_W, and ΔT_C divided by the total temperature difference ΔT. From eq. (3.81d), one can find these values with

$$\frac{\Delta T_V}{\Delta T} = \frac{z^n}{z} = z^{-(1-n)} \qquad (3.109)$$

$$\frac{\Delta T_W}{\Delta T} = \frac{B}{z} \qquad (3.110)$$

$$\frac{\Delta T_C}{\Delta T} = \frac{\sigma}{z} = 1 - \frac{B}{z} - \frac{z^n}{z} \qquad (3.111)$$

Also shown in this diagram is the variation of the local overall heat transfer coefficient k_{loc}, divided by α_{Vm} (dotted curve).

6.4 Comparison between the Approximation and the More Rigorous Analysis

The approximate calculation of the mean overall heat transfer coefficient from eq. (3.94) shall now be compared with the result of the more rigorous analysis. This can be done with the least expenditure of iterative calculations in such a way that local

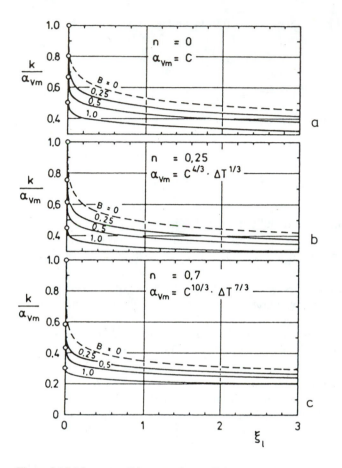

Figure 3.15 Mean overall heat transfer coefficients vs. tube length. (a) $n = 0$ (forced convection), (b) $n = 0.25$ (natural convection), (c) $n = 0.7$ (pool boiling).

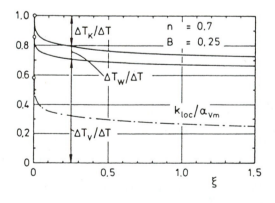

Figure 3.16 Variation of local temperature differences.

relative resistances z are given as parameter values, and the corresponding values of position ξ and mean overall resistance Z are calculated from eqs. (3.100) and (3.102) or (3.103), respectively. With the value of Z so calculated, the position ξ_{approx} can be calculated from eq. (3.94) for comparison. For $n = 0.7$, $B = 0.25$, $z_0 = 1.66996$ (from eqs. [3.104] and [3.105]), one obtains in this way the values

z	ξ	$1/Z$	$\xi_{approximation}$	$\Delta\xi/\xi$
3	0.1866	0.3713	0.1843	-1.12%
6	18.922	0.1933	18.633	-1.52%
9	168.02	0.1362	165.54	-1.48%
12	715.43	0.1035	705.41	-1.40%

The deviations are in a range of only 1 to 2% in ξ. As k/α_{Vm} vs. ξ is a relatively flat variation, the differences in k for a given position ξ are even smaller. In a diagram k/α_{Vm} vs. ξ, the curves calculated from eqs. (3.100) and (3.102) would practically coincide within the line thickness with those approximately calculated from eq. (3.94). This is shown in the range $10^{-2} \leq \xi \leq 10^3$ in Fig. 3.17 with $n = 0.7$ and $B = 0, 0.25, 0.5$, and 1.0. The figure supplements the results already given in Fig. 3.15c for $0 \leq \xi \leq 3$ and may be used to find the mean overall heat transfer coefficients for greater lengths (larger values of ξ).

In spite of the good agreement between the approximation and the more rigorous analysis, the application of the approximation eq. (3.94) has one serious disadvantage: to be able to check whether actual pool boiling exists over the whole length, the *local* heat flux $\dot{q}_{loc}$ (i.e., z not Z) or the local vaporization side temperature difference ΔT_V must be known. This shall be demonstrated in the following with a numerical example.

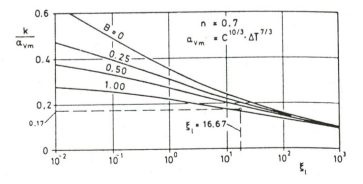

Figure 3.17 Mean overall heat transfer coefficients vs. tube length, comparison of approximate and more rigorous solutions.

6.5 Numerical Example

With steam at p_C = 2 bars (T_C = 120 °C), 98 kg/h of water are to be evaporated at p_V = 1 bar (T_V = 100 °C). What transfer surface area A is required if the heating candles are to be built from stainless steel tubing with $d_0 \times s$ = 28 × 1.5 [mm × mm] and a wall thermal conductivity of λ_w = 20 [W/(K m)]? For water at p_V = 1 bar in the range of boiling, one has $\alpha_V = c\dot{q}^n$ with c and n specified to be c = 2.0 [$W^{0.3}/(m^{0.6}$ K)] and n = 0.7 [V1]. The pool boiling range is limited by the values $\Delta T_{V,min} < \Delta T_V < \Delta T_{V,max}$ with $\Delta T_{V,min}$ ≈ 7 [K] and $\Delta T_{V,max}$ ≈ 30 [K]. At temperature differences larger than $\Delta T_{V,max}$, film boiling is reached, while, at temperature differences below $\Delta T_{V,min}$, one comes into the range of convective boiling, i.e., natural convection without bubble formation. For water of 100 °C, natural convection can be described by [V1] c = 104 [$W^{0.75}/(m^{1.5}$ K)] and n = 0.25. The boundary limit between these ranges, i.e., the beginning of bubble formation, is found from making the heat transfer coefficients equal:

$$(c\dot{q}^n)_{\text{pool boiling}} = (c\dot{q}^n)_{\text{free convection}}$$

$$\dot{q}_{min}^{0.7} = \frac{c_{fK}}{c_{Bs}}\dot{q}_{min}^{0.25}$$

$$\dot{q}_{min} = \left(\frac{104}{2}\right)^{\frac{1}{0.7-0.25}} \text{ W/m}^2$$

$$\dot{q}_{min} = 6\,506 \text{ W/m}^2$$

Corresponding to this is α_{Vm} = 934 [W/(m² K)] and $\Delta T_{V,min}$ = 6.97 [K].

Calculation of parameters

The total temperature difference $\Delta T = T_C - T_V \rightarrow \Delta T$ = 20 [K]. From eq. (3.90) follows

$$\dot{q}_{Vm} = (c\Delta T)^{\frac{1}{1-n}} = (2 \cdot 20)^{\frac{10}{3}} \text{ W/m}^2$$

$$\dot{q}_{Vm} = 218\,877 \text{ W/m}^2$$

$$\alpha_{Vm} = 10\,944 \text{ W/m}^2\text{K}$$

The characteristic length l^* from eq. (3.91), with the physical properties at $T_f \approx$ 120 °C,

$$\varrho_f = 943 \text{ kg/m}^3$$

$$\Delta h_v = 2202.9 \cdot 10^3 \text{ J/kg}$$

$$\lambda_f = 0.687 \text{ W/(K m)}$$

$$v_f = 0.244 \cdot 10^{-6} \text{ m}^2/\text{s}$$

and, with $d_o = 28$ [mm], $d_i = 25$ [mm] becomes

$$l^* = 59.99 \text{ mm}$$

The relative conduction resistance of the stainless steel wall is

$$B = \frac{\alpha_{Vm} d_o \ln(d_o/d_i)}{2\lambda_w} = 0.8682$$

If a vertical height of the tubes of $l = 1$ [m] is chosen, then follows $\xi_l = l/l^* = 16.67$. With this value, one can read from Fig. 3.17 the mean overall heat transfer coefficient divided by α_{Vm} (see broken lines in Fig. 3.17). One obtains $k \approx 1860$ [W/(m^2 K)]. The transfer surface area required becomes, with

$$\dot{Q} = \dot{M}_W \Delta h_v(T_V) = \frac{98 \text{ kg}}{3\,600 \text{ s}} 2\,257.3 \text{ kJ/kg}$$

$$\dot{Q} = 61\,449 \text{ W}$$

$$A_{req} = \frac{\dot{Q}}{k\triangle T} = 1.652 \text{ m}^2$$

$$A = n_T \pi d_o l \quad (n_T = \text{number of tubes})$$

From this, we get

$$n_{T,req} = \frac{A_{req}}{\pi d_o l} = 18.8$$

In this case, one will need at least 19 tubes of 1 [m] length each. The mean temperature difference on the vaporization side is $\Delta T_V = \Delta T \cdot k/\alpha_V$, with $\alpha_V = c(k\Delta T)^{0.7} = 3165$ [W/(m^2 K)], $\Delta T_V = 11.75$ [K], i.e., greater than $\Delta T_{V,min}$. The question, whether $\Delta T_{V,loc} (x = 1)$ is also greater than $\Delta T_{V,min}$, however, can only be checked with the rigorous calculation. From eqs. (3.100), (3.103), and (3.105), one finds:

$$z_0 = 3.052 \qquad F(z_0) = 0.4222$$

z	Θ	l / [mm]	k / [W/(m^{2}K)]
3.052	0.7155	0	3 586
4	0.6598	9.018	2 904
5	0.6170	103.9	2 416
6	0.5842	455.9	2 066
6.5	0.5703	809.8	1 927
6.702	0.5651	1 000.0	1 875

At $l = 1000.0$ [mm], a local vaporization side temperature difference of $\Delta T_{\text{V,loc}} = 0.5651 \cdot 20$ [K] $= 11.30$ [K] and a mean overall heat transfer coefficient of $k = 1875$ [W/(m² K)] are found.

If copper tubes ($\lambda_{\text{w}} = 380$ [W/(m² K)]) are chosen in place of the stainless steel tubes ($\lambda_{\text{w}} = 20$ [W/(m² K)]), the relative wall resistance from eq. (3.93) becomes only $B = 0.0457$. From this, we obtain:

$$z_0 = 1.145 \qquad F(z_0) = 0.0009865$$

z	Θ	l / [mm]	k / [W/(m²K)]
1.145	0.9601	0	9 555
3	0.7192	26.10	4 238
4	0.6598	153.0	3 254
5	0.6170	541.2	2 641
5.5985	0.5965	1 000.0	2 373

In this case, one would require only $A_{\text{req}} = 1.295$ [m²] of transfer surface, i.e., 15 copper tubes in place of the 19 stainless steel tubes.

Figure 3.18 shows, for this example, the variation of the local heat transfer coefficients $\alpha_C(x)$, $\alpha_V(x)$, and the local overall heat transfer coefficient $k_{\text{loc}}(x)$. Obviously, the larger heat transfer resistance is on the vaporization side in this case. Nevertheless, it would be wrong to conclude from this that an enhancement on the condensation side (where the resistance is much less) would not pay. Such a conclusion would be wrong because the individual heat transfer coefficients in this case are not independent of each other. Measures that contribute to make the condensate film thinner (such as longitudinally corrugated tubes [G9] or inserts which strip off the film) lead to the fact that a larger part of the total temperature difference ΔT will be

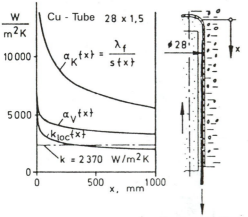

Figure 3.18 Variation of local heat transfer coefficients.

Condensate film

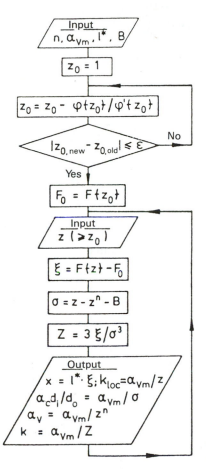

Figure 3.19 Flow diagram for the calculation procedure. ϵ = error limit.

left over on the vaporization side. Since the heat transfer coefficient α_V for pool boiling depends very strongly on the temperature difference ΔT_V ($\alpha_V \propto \Delta T_V^{7/3}$), such measures may lead to considerable savings, especially if expensive materials like tantalum or titanium are required. To check whether the assumption of laminar film condensation holds, that has been used to calculate the condensation side heat transfer coefficient, the film Reynolds number should be calculated:

$$Re_f = \frac{k\Delta T(d_o/d_i)l}{\Delta h_v \eta_f} = 104.9$$

In this range, the film flow of water is practically purely laminar still. The film thickness at $l = 1$ [m] is only $s = 124$ [μm]. For the convenient evaluation of eqs. (3.100), (3.103), and (3.105), a short computer program is used. Figure 3.19 shows a flowsheet of such a program.

7 COOLING AND PARTIAL CONDENSATION OF ONE COMPONENT FROM INERT GAS EXAMPLE: FIN-TUBE AIR COOLER AND DEHUMIDIFIER

7.1 Problem Statement

Air at a mass flowrate of 36 [kg/s], a temperature of 40 °C, and relative humidity (water vapor) of 30% is to be cooled down to 10 °C. As a coolant, a freon boiling at about 5 °C is to be used.

Questions

- How does the state of the air change?
- Will water vapor condense, and if so, how much?
- How should the apparatus be designed, and what size is required?

7.2 Calculation of the States of Air

Since the heat transfer coefficients of boiling freon are much higher than those of flowing air, externally finned tubes should be used. The resistance on the freon side can be practically neglected. For the changes of state of the air, one can assume that the temperature is $T_0 = 5$ °C at the whole tube surface. If the resistance on the freon side were not completely negligible, one could lower the pressure and, thus, the boiling temperature a little on the freon side to keep this surface temperature T_0 constant. The change of state of the air may be best represented in the Mollier-diagram (h-Y-diagram) for moist air (see e.g., [S5]). This is shown schematically in Fig. 3.20.

The change of state from the inlet conditions $T' = 40$ °C, $\varphi' = 30\%$ towards

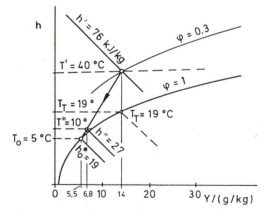

Figure 3.20 Change of state of moist air in the Mollier-diagram.

the state at the tube surface T_0, $\varphi_0 = 1$ closely approximates a linear path. As $T_0 = 5\,°C$ is far below the dewpoint of $T_T' = 19\,°C$, a condensate film will form on the surface. Moisture content Y, relative humidity φ, specific enthalpy (per unit mass of dry air) and temperature can be read from the Mollier-diagram or can be calculated. With the specific heat capacities, which may be regarded as practically constant [V1]

$$c_{pW} = c_{p,\,H_2O,\,l} = 4.190 \text{ kJ/kgK}$$

$$c_{pV} = c_{p,\,H_2O,\,g} = 1.861 \text{ kJ/kgK}$$

$$c_{pA} = c_{p,air,g} = 1.006 \text{ kJ/kgK}$$

and with the enthalpy of vaporization at the reference state

$$\Delta h_{VR} = 2\,500 \text{ kJ/kg} \qquad T_R = 0\,°C$$

the specific enthalpies are calculated according to Mollier from

$$h = c_{pA}\vartheta + Y\,(\Delta h_{VR} + c_{pV}\vartheta) \qquad (3.112)$$

Here ϑ is not, as in chapters 1 and 2, a dimensionless normalized temperature, but a temperature difference

$$\vartheta = T - T_R$$

numerically equal to the temperature in $°C$ because $T_R = 0\,°C$, which, nevertheless, should be correctly given in [K]. The moisture content Y depends on the relative humidity φ, on the saturation vapor pressure of water $p^*(\vartheta)$, the ratio of molar masses (here, $\tilde{M}_W/\tilde{M}_A = 0.622$) and the total pressure p:

$$Y = \frac{0.622\varphi p^*(\vartheta)}{p - \varphi p^*(\vartheta)} \qquad (3.113)$$

With a suitable equation for the equilibrium vapor pressure, one obtains from this at the total pressure of $p = 10^5$ Pa ($= 1$ bar):

State of the air at the inlet

$$T' = 40\,°C \qquad \varphi' = 0.3 \qquad p^*(40\,°C) = 7.38 \cdot 10^3 \text{ Pa}$$

$$(h,\,Y)' = (76.50 \text{ kJ/kg}, \quad 14.08 \text{ g/kg})$$

at the phase interface (the conduction resistance of the thin condensate film is neglected)

$$T_0 \approx 5° \qquad \varphi_0 = 1.0 \qquad p^*(5°C) = 0.872 \cdot 10^{-3} \text{ Pa}$$

$$\boxed{(h, Y)_0 = (18.76 \text{ kJ/kg}, \quad 5.47 \text{ g/kg})}$$

at the outlet (linear variation, $Le \approx 1$)

$$T'' = 10°C$$

$$\frac{h'' - h_0}{h' - h_0} = \frac{Y'' - Y_0}{Y' - Y_0} \qquad h'' = c_{pL}\vartheta'' + Y''(\Delta h_{VR} + c_{pv}\vartheta'')$$

$$Y'' = \frac{Y_0 + (c_{pA}\vartheta'' - h_0)\Delta Y / \Delta h}{1 - (\Delta h_{VR} + c_{pv}\vartheta'')\Delta Y / \Delta h} \tag{3.114}$$

$$\frac{\Delta Y}{\Delta h} = \frac{Y' - Y_0}{h' - h_0} = 0.1491 \ (10^{-3} \text{ kg})/\text{kJ}$$

$$\boxed{(h, Y)'' = (26.89 \text{ kJ/kg}, \quad 6.68 \text{ g/kg})}$$

7.3 Energy and Mass Balances, Surface Area

a. Balance volume—air side without the condensate film:

$$\frac{\dot{M}}{1 + Y'}(h' - h'') - \dot{M}_C h_{v0} - \dot{Q}_0 = 0 \tag{3.115}$$

b. Balance volume—air side including the condensate film:

$$\frac{\dot{M}}{1 + Y'}(h' - h'') - \dot{M}_C h_{w0} - \dot{Q}_W = 0 \tag{3.116}$$

The third possible balance around the condensate film has to coincide, of course, with the difference of the two eqs. (3.115) and (3.116). Here one has to distinguish carefully between the heat flow rate $\dot{Q}_0$ transferred to the film surface and $\dot{Q}_W$ transferred to the wall (and thus to the vaporizing freon):

$$\dot{Q}_W = \dot{Q}_0 + \dot{M}_C \, \Delta h_{v0} \tag{3.117}$$

In the balance equations, $\dot{M}$ is the total mass flow rate of moist air at the inlet ($\dot{M} = 36$ kg/s), and $\dot{M}/(1 + Y')$ is the fraction of dry air, i.e., the carrier stream remaining constant. The condensate rate $\dot{M}_C$ follows from the air side mass balance:

$$\frac{\dot{M}}{1 + Y'}(Y' - Y'') - \dot{M}_C = 0 \tag{3.118}$$

Together with the values given or already calculated, we get

$$\dot{M}_C = 0.2627 \ [\text{kg/s}]$$

The specific enthalpy of water vapor at the phase interface h_{v0} follows from

$$h_{v\,0} = \Delta h_{vR} + c_{pv}\vartheta_0 \tag{3.119}$$

$$h_{v\,0} = 2\,509 \ \text{kJ/kg}$$

That of liquid water at ϑ_0 is

$$h_{w0} = c_{pw}\vartheta_0 \tag{3.120}$$

$$\boxed{h_{w0} = 20.95 \ \text{kJ/kg}}$$

The difference between these two specific enthalpies is

$$\Delta h_{v0} = h_{v0} - h_{w0} = 2488 \ [\text{kJ/kg}]$$

From eqs. (3.115) and (3.166) or (3.117) follows:

$$\dot{Q}_0 = [35.5(76.50 - 26.89) - 0.2627 \cdot 2509] \ [\text{kW}]$$

$$\dot{Q}_0 = (1761.2 - 659.1) \ [\text{kW}] = 1.102 \ [\text{MW}]$$

$$\dot{Q}_w = (1.102 + 0.654) \ [\text{kW}] = 1.756 \ [\text{MW}]$$

The transfer surface area required can be found either from the condensation rate and an equation for the kinetics of mass transfer, or from the heat rate $\dot{Q}_0$ from gas to film surface with an equation for the kinetics of heat transfer. The reason that only one of the equations of kinetics is needed here is that the outlet state has been calculated using an assumption on the ratio of the two transport phenomena (Lewis law, with the Lewis number $Le \approx 1$; see [S2]). As the temperature of the film surface is constant, the logarithmic mean temperature difference can be applied:

$$A_{req} = \frac{\dot{Q}_0}{\alpha_{g0}\,\Delta T_{LM}} \tag{3.121}$$

With an estimated value of $\alpha_g^{(0)} = 50 \ [\text{W/(m}^2\text{ K)}]$ and

$$\Delta T_{LM} = \frac{(T' - T_0) - (T'' - T_0)}{\ln[(T' - T_0)/(T'' - T_0)]} = \frac{35 - 5}{\ln(35/5)} \ \text{K}$$

$$\Delta T_{LM} = 15.4 \ \text{K}$$

the required transfer surface area becomes $A_{req}^{(0)} \approx 1430 \ [\text{m}^2]$.

7.4 Choice of Dimensions of Tubes and Flow Cross Section

Finned tubes as shown in Fig. 3.21 are to be used. The fins may be manufactured as continuous sheets for a whole row of tubes. On the air side, therefore, small channels of gap width 4.5 [mm] are created. With $\rho \approx 1.2$ [kg/m^3], i.e., $\dot{V} \approx 30$ [m^3/s], the flow velocity will be $w = 6$ [m/s] for a flow cross section of $S \approx 5$ [m^2] (see eq. [3.18] for comparison). This is the velocity that would be found in the free cross section, the superficial velocity. As about half of it is blocked by tube and fin in the narrowest cross section, a velocity around 12 [m/s] is to be expected at the tubes. The flow cross section can be attained, e.g., with 30 tubes per row → height $H = (30 \times 60)$ [mm] $= 1800$ [mm], and a length of $L = 2750$ [mm] (→ $H \times L = 4.95$ [m^2]).

The surface of such a row of tubes is then

$$A_{Rt,0} = \pi d_o L n_R = \pi \cdot 25 \cdot 10^{-3} \cdot 2.75 \cdot 30^2$$

$$A_{Rt,0} = 6.480 \text{ m}^2 \text{ (without fins)}$$

The surface of one fin (see Fig. 3.21):

$$A_{fin} = 2[(2s)^2] - \pi r_0^2$$
$$= 2[60^2 - \pi(12.5)^2] \text{ mm}^2 = 6\,218 \text{ mm}^2$$

Surface enlargement ration:

$$\frac{A}{A_0} = \frac{A_{fin} + \pi d_o(s_t - \delta)}{\pi d_o s_t} \tag{3.122}$$

$$\frac{A}{A_0} = 15,83 + \frac{4.5}{5} = 16,73$$

$$A_{Rt} = 16.73 \cdot 6.48 \text{ m}^2 = 108.4 \text{ m}^2$$

With the first estimation for α_{g0} (and a fin efficiency of one), this would mean that about 14 of these rows of finned tubes would be needed.

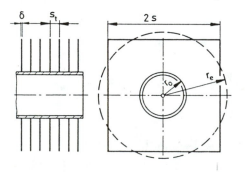

Figure 3.21 Dimensions of the finned tubes.

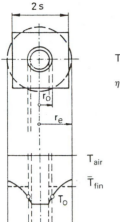

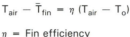

$$T_{air} - \bar{T}_{fin} = \eta \, (T_{air} - T_o)$$

η = Fin efficiency

Figure 3.22 Temperature variation in the fin.

7.5 Calculation of Fin Efficiency

The fin surface cannot be inserted into the calculation of transfer surface area with the total temperature difference $T_{air} - T_0$, since, with increasing distance from the base, the fin temperature deviates more and more from the tube base temperature T_0 due to the radial conduction in the fin and the transfer from the air to the fin surface, as schematically shown in Fig. 3.22. A fin efficiency is defined

$$\eta_{fin} = \frac{T_{air} - \bar{T}_{fin}}{T_{air} - T_0} \qquad (3.123)$$

accounting for this diminution of the driving temperature difference. Fin efficiencies can be approximately calculated under the assumptions of one dimensional heat conduction ("thin" fins) and constant heat transfer coefficients at the whole transfer surface (see, e.g., Kern and Kraus [K2]).

For annular fins, one obtains the solution shown in Fig. 3.23 for fin efficiency η as a function of the "fin number" Φ and the ratio ρ of base radius to edge radius of the annular fins. In the limit $\rho \to 1$ (plane fin with $(r_e - r_0) = l =$ fin height), the solution becomes especially simple:

$$\eta_{fin} \atop {\varrho \to 1} = \frac{\tanh \Phi}{\Phi} \qquad (3.124)$$

For radius ratios less than one, the fin efficiency can be calculated from the equation given in Fig. 3.23 in terms of the cylindrical functions I_0, I_1, K_0, K_1 (modified Bessel's and McDonald's functions [A1]), or simply read from the diagram. The quadratic fins are treated approximately as circular fins of having the same area with the edge radius r_e (see Fig. 3.21):

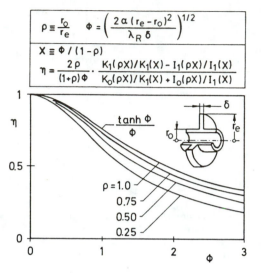

Figure 3.23 Fin efficiency as a function of dimensionless fin height Φ and radius ratio ρ.

$$\pi r_e^2 = (2s)^2 \qquad r_e = \frac{2}{\sqrt{\pi}} s \qquad (3.125)$$

and from this

$$\boxed{\varrho = \frac{r_0}{r_e}} = \frac{12.5\sqrt{\pi}}{60} \boxed{= 0.369}$$

The "fin number" (dimensionless fin height)

$$\boxed{\Phi = \left(\frac{2\alpha(r_e - r_0)^2}{\lambda_{\text{fin}}\,\delta} \right)^{\frac{1}{2}}} \qquad (3.126)$$

with

$$\frac{\lambda_{\text{fin}}\,\delta}{2(r_e - r_0)^2} = \frac{200 \cdot 0.5 \cdot 10^{-3}}{2(21.35)^2 10^{-6}}\ \text{W/(m}^2\text{K)} \approx 110\ \text{W/(m}^2\text{K)}$$

$$\lambda_{\text{Aluminum}} = 200\ \text{W/(m}^2\text{K)}$$

and

$$\alpha = \alpha_g^{(0)} = 50\ \text{W/(m}^2\text{K)}$$

becomes

$$\Phi^{(0)} = 0.675$$

The fin efficiency is, therefore,

$$\eta(0.675; \ 0.369) \approx 0.80.$$

7.6 Calculation of Heat Transfer Coefficient

For the calculation of mean heat transfer coefficients, a correlation by Schmidt [S6] based on experimental data with bundles of finned tubes of various dimensions (in-line arrangement) is recommended in HEDH [H3, p. 2.5.3–11]:

$$Nu_d = 0.30 \, Re_d^{0.625} \left(\frac{A}{A_0} \right)^{-0.375} Pr^{1/3} \qquad (3.127)$$

The maximum velocity, i.e., the volume flowrate divided by the narrowest flow cross section is to be used as the characteristic velocity in the Reynolds number, and the tube diameter d_0 is the characteristic length in the dimensionless numbers Nu_d and Re_d.

The smallest open fraction of the cross section can be calculated from Fig. 3.21 to be

$$\psi_{min} = 1 - \frac{s_t d_a + (2s - d_o)\delta}{s_t 2s}$$

$$\psi_{min} = \left(1 - \frac{\delta}{s_t} \right) \left(1 - \frac{d_o}{2s} \right) \qquad (3.128)$$

$$\psi_{min} = 0.525$$

Physical properties at $T_0 + \Delta T_{LM} \approx 20\,°C$:

$Pr = 0.7 \qquad\qquad v = 15 \cdot 10^{-6} \ m^2/s \qquad\qquad \lambda = 0.026 \ W/(mK)$

$\varrho = 1.19 \ kg/m^3 \qquad w_{max} = 11.64 \ m/s \qquad \rightarrow Re_d = w_{max} d_o/v = 19\,400$

$Nu_d = 0.3 \cdot 19\,400^{0.625}(16,73)^{-0.375} 0.7^{1/3} = 44,32$

$\alpha_g = (\lambda/d_o)Nu_d$

$$\boxed{\alpha_g = 46.1 \ W/(m^2 K)}$$

As the fins are not charged with the outer heat flow rate $\dot{Q}_0$ (gas-to-film surface), but with the greater heat flow rate $\dot{Q}_w$, the fin efficiency η has to be calculated with an effective heat transfer coefficient α_g^* enlarged by the ratio $\dot{Q}_w/\dot{Q}_0$:

$$\Phi = \left(\frac{2\alpha_g^*(r_e - r_0)^2}{\lambda_{fin}\delta} \right)^{\frac{1}{2}} \qquad (3.129)$$

$$\alpha_g^* = \alpha_g \frac{\dot{Q}_W}{\dot{Q}_0} \tag{3.130}$$

$$\alpha_g^* = 46.1 \cdot 1.586 \ \text{W}/(\text{m}^2\text{K}) = 73.1 \ \text{W}/(\text{m}^2\text{K})$$

$$\Phi = \left(\frac{73.1}{110}\right)^{\frac{1}{2}} = 0.815$$

$$\eta(0.815; \ 0,369) = 0.745$$

$$\rightarrow \quad \dot{Q}_0 = \alpha_g A_0 \left(1 - \frac{\delta}{s_t} + \frac{\eta A_{\text{fin}}}{A_0}\right) \Delta T_{\text{LM}}$$

$$\boxed{A_{0,\text{req}}} = \frac{\dot{Q}_0}{\alpha_g[(1 - \delta/s_t) + \eta A_{\text{fin}}/A_0]\Delta T_{\text{LM}}} \tag{3.131}$$

$$\boxed{A_{0,\text{req}}} = \frac{1.107 \cdot 10^6}{46.1 \cdot 12.69 \cdot 15.4} \ \text{m}^2 \ \boxed{= 123 \ \text{m}^2}$$

$$\frac{A_{0,\text{req}}}{A_{\text{Rt},0}} = \frac{123 \ \text{m}^2}{6.48 \ \text{m}^2} \approx 19$$

i.e., 19 rows of tubes will be needed. The dimensions of the whole block (see Fig. 3.24) are, therefore, $L \times H \times B = 2750 \times 1800 \times 1140 \ [\text{mm}^3]$. The volume is $V = 5.64 \ [\text{m}^3]$.

7.7 Recalculation of the Heat Transfer Coefficient by a Different Method

In place of eq. (3.127) for bundles of finned tubes, the plane channels between the fins could also be regarded as parallel plate ducts (see eq. [1.140] with $f = 0.86$). As

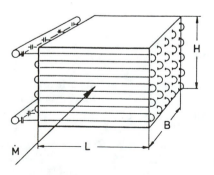

Figure 3.24 Fin-tube air cooler, outer dimensions.

a characteristic flow velocity, one should use not the maximum but rather a mean velocity in this case (to be calculated from a volumetric void fraction):

$$\psi = \left(1 - \frac{\delta}{s_t}\right)\left[1 - \frac{\pi}{4}\left(\frac{d_o}{2s}\right)^2\right]$$

(3.132)

$$\psi = 0.777 \qquad\qquad \longrightarrow \quad w = 7.87 \text{ m/s}$$

$$d_h = 2(s_t - \delta) \qquad\qquad \longrightarrow \quad d_h = 9 \text{ mm}$$

$$Re = \frac{wd_h}{v} = 4719 \qquad \longrightarrow \quad \text{Eq. (1.140) with } f = 0.86$$

$$\frac{d_h}{L} = \frac{9}{60}, \qquad\qquad\qquad Nu = 17.36$$

$$\alpha_{g,gap} = 50.2 \text{ W/(m}^2\text{K)}$$

This value is only 9% higher than the one calculated from eq. (3.127).

CHAPTER FOUR

IMPORTANT DATA FOR THERMAL DESIGN OF HEAT EXCHANGERS

In the following, we will give only a few numerical values and simple formulas for a first estimation of the most important data for thermal design of heat exchangers. For more accurate data and calculation procedures, see VDI Wärmeatlas [V1, pp. Da–Df], HEDH [H3, volume 5] or Reid, Prausnitz, and Sherwood [R1].

1 HEAT CAPACITIES

The energy storage capacity of a body is expressed by the amount of energy that has to be transferred to it relative to its amount of substance or its mass to raise its temperature by one degree under certain specified experimental conditions. As the experimental conditions to be kept constant, one can choose constant volume or constant pressure. At constant volume, the energy added can only be stored in the form of internal energy, i.e., as the energy of molecular and atomic motion. At constant pressure, expansion work is done on the surroundings additionally. One defines

$$c_v \equiv \left(\frac{\partial u}{\partial T} \right)_v$$

$$c_p \equiv \left(\frac{\partial h}{\partial T} \right)_p$$

1.1 Gases

The heat capacities of ideal gases, according to the kinetic theory of gases, depend on the number of degrees of freedom of translation, rotation, and vibration. For monoatomic gases with three degrees of freedom of translation, the molar specific heat capacity at constant volume becomes

$$\tilde{c}_{v,1A} = \frac{3}{2}\tilde{R} \approx 12.5 \text{ J/(mol K)} \tag{4.1}$$

and with $\tilde{c}_p = \tilde{c}_v + \tilde{R}$

$$\tilde{c}_{p,1A} = \frac{5}{2}\tilde{R} \approx 20.8 \text{ J/(mol K)} \tag{4.2}$$

at constant pressure.

For diatomic and triatomic gases, two or three degrees of freedom of rotation are added, so that one obtains

$$\tilde{c}_{p,2A} = \frac{7}{2}\tilde{R} \approx 29.1 \text{ J/(mol K)} \tag{4.3}$$

$$\tilde{c}_{p,3A} = 4\tilde{R} \approx 33.3 \text{ J/(mol K)} \tag{4.4}$$

Since the rotational degrees of freedom are only excited at higher temperatures, the heat capacities of these gases are temperature dependent. In practice, one will often need the mass specific (c_p) or the volumetric heat capacities (ρc_p). Because of $\rho = n\tilde{M}$ and $c = \tilde{c}/\tilde{M}$, one finds for the volumetric heat capacity

$$\varrho c_p = n\tilde{c}_p \tag{4.5}$$

For ideal gases, we can replace n by $p/(\tilde{R}\,T)$, so that the volumetric heat capacity of gases can be easily estimated from the simple relation

$$\boxed{(\varrho c_p)_g \approx (2.5\ldots5)\frac{p}{T}} \tag{4.6}$$

At $p = 10^5$ [Pa], $T = 298$ [K], one obtains for diatomic gases (numerical factor = 3.5)

$$(\varrho c_p)_{2A\text{-Gas}}(1 \text{ bar}, 25\,^\circ\text{C}) \approx 1.2 \cdot 10^3 \text{ J/(m}^3\text{K)}$$

1.2 Liquids

The volumetric heat capacities of liquids are usually in the range

$$\boxed{10^6 \text{ J/(m}^3\text{K)} < (\varrho c_p)_l < 4.2 \cdot 10^6 \text{ J/(m}^3\text{K)}} \tag{4.7}$$

i.e., three orders of magnitude higher than for gases at ambient conditions. Most organic liquids have values in the range 1 ... 2 [MJ/(m³ K)] and water has the relatively high value of 4.2 [MJ/(m³K)].

1.3 Solids

All compact, i.e., non-porous, solids have volumetric heat capacities lying in the same range as those for liquids:

$$\boxed{10^6 \text{ J/(m}^3\text{K}) < (\varrho c_p)_s < 4 \cdot 10^6 \text{ J/(m}^3\text{K})}$$ (4.8)

2 THERMAL CONDUCTIVITIES

The thermal conductivity λ is defined by Fourier's law

$$\lambda \equiv -\dot{q}/\text{grad } T$$ (4.9)

For some solid, liquid, and gaseous substances, it is plotted vs. temperature in Fig. 4.1. Characteristic values for a few materials at 20 °C and 1 bar are

Material	$\lambda/(\text{W/Km})$	Material	$\lambda/(\text{W/Km})$
Silver	460	Glass, Ceramic	1...3
Copper	380	Water	0.6
Aluminum	200	H_2	0.18
Stainless steel	12...20	Air	0.026

3 HEAT TRANSFER COEFFICIENTS

Heat transfer coefficients are defined by the linear relation for the kinetics of heat transfer, which is sometimes referred to as "Newton's law of cooling"

$$\alpha \equiv \frac{\dot{Q}_{12}}{A(T_1 - T_2)}$$ (4.10)

corresponding to the integral form of Fourier's law. They can, therefore, be reduced to the conductivity λ and a length s ($\rightarrow$ thickness of the thermal boundary layer) by the expression

$$\alpha = \lambda/s$$ (4.11)

For flowing media, s, in turn, depends on λ, ρc_p, η, the flow velocity w, and the geometric parameters of the fluid dynamic problem. These dependencies are usually written in dimensionless form as

$$Nu = Nu\left(Re, Pr, \frac{d}{l}, \dots\right)$$ (4.12)

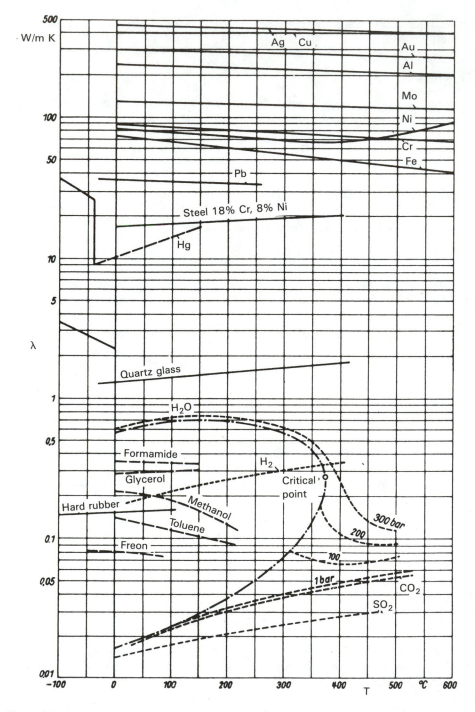

Figure 4.1 Heat conductivity of various solid, liquid and gaseous substances vs. temperature (VDI–WA).

Formulas of this type are found at various places in the text. For

- duct flow
  tube and annulus: eqs. (1.128) and (1.140), in plate heat exchangers: eq. (2.122), in spiral-plate heat exchangers: eq. (2.132)
- single bodies in cross flow: eq. (3.51)
- fixed beds: eq. (3.50)
- bundles of finned tubes: eq. (3.127)

For other cases, see HEDH and VDI-WA.

A first approximation for heat transfer coefficients in turbulent forced flow may be obtained from the formula

$$\alpha_{\text{forced convection}} \approx 0.004 \, \varrho c_p w \left(\frac{\lambda}{\eta \, c_p} \right)^{2/3} \qquad (4.13)$$

For air with $(\eta c_p / \lambda) = Pr = 0.7$, $\rho c_p \approx 1000$ [J/(m^{3}K)] and velocities in the range of 1 to 50 [m/s] (see chapter 3, section 1.9), one obtains

$$\alpha_{\text{air}} \approx 5 \text{ to } 250 \text{ [W/(m}^2 \text{ K)]}$$

For water with $Pr \approx 2$ (at 90 °C), $\rho c_p \approx 4.2 \cdot 10^6$ [J/(m^3 K)], and flow velocities of 0.2 to 2 [m/s] (see chapter 3, section 1.9), we find

$$\alpha_{\text{water}} \approx 2\,000 \text{ to } 20\,000 \text{ [W/(m}^2 \text{ K)]}$$

Note: The factor 0.004 in eq. (4.13) is not a true constant. It depends, in reality, on flow velocity w and on geometric parameters. The heat transfer coefficient does not increase linearly with w, but with a smaller power: $\alpha \propto w^n$, n $\approx 0.6 \ldots 0.9$. Equation (4.13) should, therefore, only be used as a first approximation for α! It gives the correct order of magnitude for various fluids.

For liquid films flowing downward under the effect of gravity (e.g., during condensation of pure vapors) the order of magnitude of the heat transfer coefficient can be estimated from the equation

$$\alpha_{\text{Film}} \approx (0.1 \ldots 1)\lambda_f \left(\frac{g}{v_f^2} \right)^{1/3} \qquad (4.14)$$

For water with $\lambda_f \approx 0.6$ [W/(K m)] and $v_f \approx 10^{-6}$ [m^2/s] (at 20 °C), one finds

$$\alpha_{\text{water film}} \approx 2\,600 \ldots 26\,000 \text{ [W/(m}^2 \text{ K)]}$$

CROSSFLOW OVER n ROWS OF TUBES

CALCULATION OF TEMPERATURE VARIATION
AND MEAN OUTLET TEMPERATURE

With ϑ_j for the temperature of stream 1 in the tubes of row j, Θ_j for the temperature of stream 2 flowing over the tubes perpendicular to the tube axes, after the row j (see eqs. [2.71] and [2.72]); and the dimensionless length coordinate $x = \zeta_1/\varphi(N_2/n)$ in flow direction of stream 1 through the tubes, the eqs. (2.34) and (2.35) are brought into the most compact form:

$$-\frac{d\vartheta_j}{dx} = \vartheta_j - \Theta_{j-1} \tag{A1}$$

$$\boxed{e^x\vartheta_j = \vartheta_j(0) + \int_0^x e^x \Theta_{j-1}\, dx} \tag{A2}$$

The temperature Θ_{j-1} of medium 2 after the row $(j - 1)$ can be calculated from eq. (2.40) as a function of x, if ϑ_{j-1} and Θ_{j-2} are known as functions of x:

$$\boxed{\Theta_{j-1} = a\Theta_{j-2} + b\vartheta_{j-1}} \tag{A3}$$

183

$$a \equiv e^{-N_2/n}, \qquad\qquad b \equiv \frac{N_2/n}{\varphi(N_2/n)} = 1 - a \qquad (A4)$$

$$0 \le x \le B, \qquad\qquad B \equiv \frac{bn}{R} \qquad (A5)$$

$$R = \frac{N_2}{N_1} = \frac{\dot{M}_1 c_{p1}}{\dot{M}_2 c_{p2}}$$

Starting with the value $\Theta_0 = 0$ (inlet temperature of medium 2) and alternately applying eqs. (A2) and (A3), one obtains the temperature variation in successive rows of tubes ($\vartheta_j(0) = 1$, inlet temperature of medium 1):

$$e^x \vartheta_1 = 1 \qquad\qquad e^x \Theta_1 = b$$

$$e^x \vartheta_2 = 1 + bx \qquad\qquad e^x \Theta_2 = (a+1)b + b^2 x$$

$$e^x \vartheta_3 = 1 + (a+1)bx + b^2\frac{x^2}{2} \qquad e^x \Theta_3 = (a^2 + a + 1)b + (2a+1)b^2 x + b^3\frac{x^2}{2}$$

$$e^x \vartheta_4 = 1 + (a^2 + a + 1)bx + (2a+1)b^2\frac{x^2}{2} + b^3\frac{x^3}{6} \qquad\qquad (A6)$$

$$\vdots$$

$$e^x \vartheta_6 = 1 + (a^4 + a^3 + a^2 + a + 1)bx + (4a^3 + 3a^2 + 2a + 1)b^2\frac{x^2}{2!} +$$

$$+ (6a^2 + 3a + 1)b^3\frac{x^3}{3!} + (4a+1)b^4\frac{x^4}{4!} + b^5\frac{x^5}{5!}$$

The mean outlet temperature from n rows of tubes is calculated from

$$\vartheta_n'' = \frac{1}{n}\left[\sum_{j=1}^n \vartheta_j(B)\right] \qquad (A7)$$

Via an overall balance, this can also be calculated from the integral average temperature of stream 2:

$$R(1 - \vartheta_n'') = \frac{1}{B}\int_0^B \Theta_n(x)\, dx \qquad (A8)$$

To calculate the maximum efficiencies at a fixed capacity rate ratio R, one now considers the case $N_2/n \to \infty$, which results in simple numbers for the quantities a, b, B:

$$\lim_{N_2/n \to \infty}\begin{pmatrix} a \\ b \\ c \end{pmatrix} = \begin{pmatrix} a_\infty = 0 \\ b_\infty = 1 \\ B_\infty = n/R \end{pmatrix} \qquad (A9)$$

From eq. (A3), we find in this limit

$$\Theta_{j-1}(x) = \vartheta_{j-1}(x) \tag{A10}$$

i.e., stream 2 leaves each row of tubes in local thermal equilibrium with the corresponding local temperature of stream 1 in these tubes.

The temperature variations $\vartheta_j(x)$ are then simply

$$
\begin{aligned}
e^x \vartheta_1 &= 1 \\
e^x \vartheta_2 &= 1 + x \\
e^x \vartheta_3 &= 1 + x + \frac{x^2}{2!} \\
e^x \vartheta_4 &= 1 + x + \frac{x^2}{2!} + \frac{x^3}{3!} \\
e^x \vartheta_j &= 1 + x + \frac{x^2}{2!} + \frac{x^3}{3!} + \cdots + \frac{x^{j-1}}{(j-1)!} \\
e^x \vartheta_n &= 1 + x + \frac{x^2}{2!} + \frac{x^3}{3!} + \dots\dots\dots + \frac{x^{n-1}}{(n-1)!}
\end{aligned}
\tag{A11}
$$

In this case, the progression of the power series becomes immediately obvious and the validity of the solution

$$\vartheta_j = \sum_{m=0}^{j-1} \frac{x^m}{m!} e^{-x} \tag{A12}$$

can be proven for arbitrary j at $\Theta_{j-1} = \vartheta_{j-1}$ by inserting it into the differential eq. (A1). The mean outlet temperature is again obtained from eq. (A7) with $B_\infty = n/R$:

$$\vartheta''_{n,\infty} = \frac{1}{n} \sum_{j=1}^{n} \sum_{m=0}^{j-1} \frac{(n/R)^m}{m!} e^{-n/R} \tag{A13}$$

This sum can be rewritten by rearranging its terms in the form

$$\vartheta''_{n,\infty} = \frac{(n/R)^{n-1}}{(n-1)!} e^{-n/R} - \left(\frac{1}{R} - 1 \right) \sum_{m=0}^{n-2} \frac{(n/R)^m}{m!} e^{-n/R} \tag{A14}$$

At equal capacities ($R = 1$), the mean outlet temperature is just equal to the last term in the sum of eq. (A11), which can be seen to be the case for $x \to n$. For an arbitrary number of rows, the maximum efficiency at $R = 1$ is given by

$$\boxed{\varepsilon_{\infty,n} = 1 - \frac{n^n}{n!} e^{-n}} \tag{A15}$$

and for large n (practically, for $n > 5$), using Stirling's formula, by

$$\varepsilon_{\infty,n} \approx 1 - (2\pi n)^{-1/2} \tag{A16}$$

The general progression of the series in eq. (A6) may be recognized, too, by regarding the coefficients $A_{j,m}$ of the mth power of x in the row j:

j	A_{j1}	A_{j2}	A_{j3}	A_{j4}
1	0			
2	1			
3	$1+a$	1		
4	$1+a+a^2$	$1+2a$	1	
5	$1+a+a^2+a^3$	$\boxed{1+2a+3a^2}$	$\boxed{1+3a}$	1
6	$1+a+a^2+a^3+a^4$	$1+2a+3a^2+4a^3$	$\boxed{1+3a+6a^2}$	$1+4a$

These coefficients $A_{j,m}$ are, in turn, power series in a, whose coefficients form the sequences $1,1,1,1 \ldots$, $1,2,3,4,5 \ldots$, $1,3,6,10 \ldots$, $1,4,10 \ldots$, and so on. These sequences may be recognized as binomial coefficients by using Pascal's triangle, e.g.,

$$A_{j,m} = \sum_{k=0}^{j-m-1} \binom{m-1+k}{k} a^k \tag{A17}$$

This can be explained by the generation of these coefficients from the addition of two corresponding coefficients of the previous rows. With this, the general solution becomes

$$\vartheta_j(x) = \sum_{m=0}^{j-1} A_{jm} \frac{(bx)^m}{m!} e^{-x} \tag{A18}$$

$$\Theta_{j-1}(x) = \sum_{m=1}^{j-1} A_{jm} \frac{(bx)^{m-1}}{(m-1)!} e^{-x} \tag{A19}$$

If we now apply eq. (A3) to these two general expressions for ϑ_j and Θ_{j-1}, a recursion formula is obtained for the coefficients

$$A_{jm} = A_{j-1,m-1} + a \cdot A_{j-1,m} \tag{A20}$$

This condition is, in fact, fulfilled by the coefficients from eq. (A17) as may be seen from the table (see, e.g., $A_{63} = A_{52} + a \cdot A_{53}$). The mean outlet temperature is eventually obtained from the mixture of the n streams:

$$\vartheta_n'' = \frac{1}{n} \sum_{j=1}^{n} \vartheta_j(B) \tag{A21}$$

The solution of this problem was already given, in a slightly different form in 1968 in the Dr.-Ing. dissertation (Ph.D. thesis) of Schedwill [S1] and presented in diagrams $\epsilon(N, R)$ for row numbers one to three (there denoted as operation characteristics $\Phi(kF/W_1, W_1/W_2)$). Schedwill's calculations have also been the basis of the diagrams for ϵ_1 vs. ϵ_2, with Θ and N_1, N_2 as parameters given by Roetzel in the VDI–WA [V1] for crossflow with one, two, three, and four rows of tubes. Schedwill [S1] has also shown that his solution tends to Nusselt's series solution for ideal crossflow (see eqs. [2.51], [2.52], and [2.56]) in the limit of infinite numbers of rows.

TEMPERATURE VARIATION AND EFFICIENCY OF A HEAT EXCHANGER WITH THREE PASSES IN A LATERALLY MIXED SHELL-SIDE STREAM

Special case. Equal capacities of both streams.

Assumption: First and third pass of the Y-stream in counterflow, second pass in parallel flow to the laterally mixed shell-side stream X (see Figs. 2.53, $n = 3$, and 2.54 [3, 1, 2]).

Normalized temperatures of the three internal passes: Θ_1, Θ_2, Θ_3
Normalized temperature of the shell-side stream: Θ_0
Dimensionless length coordinate: $0 \leq \zeta \leq (Y/3)$
$Y = kA/(\dot{M}c_p)_Y$, $X = kA/(\dot{M}c_p)_X$, $C = X/Y$, special case: $C = -1$.

Differential equations:

$$\frac{d\Theta_0}{d\zeta} = -C(3\Theta_0 - \Theta_1 - \Theta_2 - \Theta_3) \tag{B1}$$

$$-\frac{d\Theta_1}{d\zeta} = \Theta_1 - \Theta_0 \tag{B2}$$

$$\frac{d\Theta_2}{d\zeta} = \Theta_2 - \Theta_0 \tag{B3}$$

$$-\frac{d\Theta_3}{d\zeta} = \Theta_3 - \Theta_0 \tag{B4}$$

or in matrix notation

$$\boxed{\frac{d\Theta_i}{d\zeta} = [A]\,\Theta_i}$$

with

$$[A] = \begin{pmatrix} -3C & +C & +C & +C \\ +1 & -1 & & \\ -1 & & +1 & \\ +1 & & & -1 \end{pmatrix}$$

The eigenvalues r_i are found from this with eq. (2.99) as the roots of the characteristic equation

$$\boxed{r(1+r)[r^2 - 3Cr - (1+C)] = 0} \tag{B5}$$

to

$$r_1 = -\frac{3C}{2} + \left[\left(\frac{3C}{2}\right)^2 + (1+C)\right]^{1/2} \tag{B6}$$

$$r_2 = -\frac{3C}{2} - \left[\left(\frac{3C}{2}\right)^2 + (1+C)\right]^{1/2} \tag{B7}$$

$$r_3 = -1 \tag{B8}$$

$$r_4 = 0 \tag{B9}$$

In the special case, $C = -1$ follows $r_1 = 3$, $r_2 = r_4 = 0$ (double root), $r_3 = -1$. The general solution for $C = -1$, therefore, reads

$$\Theta_i = A_i + B_i\zeta + C_i\,e^{-\zeta} + D_i\,e^{3\zeta} \qquad (i = 0, 1, 2, 3) \tag{B10}$$

and its derivative

$$\frac{d\Theta_i}{d\zeta} = B_i - C_i e^{-\zeta} + 3D_i e^{3\zeta} \qquad (i = 0, 1, 2, 3) \tag{B11}$$

The constants are determined using the differential eqs. (B1)–(B4) and the boundary and coupling conditions ($X = -Y = N$ for $C = -1$):

$$\Theta_0(N/3) = 0 \tag{B12}$$

$$\Theta_1(0) - 1 = 0 \tag{B13}$$

$$\Theta_1(N/3) - \Theta_2(N/3) = 0 \tag{B14}$$

$$\Theta_2(0) - \Theta_3(0) = 0 \tag{B15}$$

From the differential equations, one finds, first

$$A_1 = A_0 - B_0 \qquad B_1 = B_0 \qquad C_0 = 0 \qquad D_1 = D_0/4 \tag{B16}$$

$$A_2 = A_0 + B_0 \qquad B_2 = B_0 \qquad C_2 = 0 \qquad D_2 = -D_0/2 \tag{B17}$$

$$A_3 = A_0 - B_0 \qquad B_3 = B_0 \qquad C_3 = -C_1 \qquad D_3 = D_0/4 \tag{B18}$$

Then, from the boundary and coupling conditions

$$A_0 + \frac{N}{3} B_0 \qquad\qquad + \quad e^N D_0 = 0 \tag{B19}$$

$$A_0 - \quad B_0 + \quad C_1 + \quad \frac{1}{4} D_0 = 1 \tag{B20}$$

$$- \quad 2B_0 + e^{-N/3} C_1 + \frac{3}{4} e^N D_0 = 0 \tag{B21}$$

$$2B_0 + \quad C_1 + \frac{3}{4} D_0 = 0 \tag{B22}$$

From these, the constants are found with $x = e^{-N/3}$ to be

$$A_0 = \frac{(1 - x^4)N + 8(1 + x)}{\Phi} \tag{B23}$$

$$B_0 = \frac{-3(1 - x^4)}{\Phi} \tag{B23}$$

$$C_1 = \frac{6(1 - x^3)}{\Phi} \tag{B25}$$

$$D_0 = \frac{-8x^3(1 - x)}{\Phi} \tag{B26}$$

With the common denominator

$$\Phi = (N + 3)(1 + x^4) + 6(1 - x^3) + (1 + x)(8 - 2x^3) \tag{B27}$$

The normalized change in temperature $\epsilon = \Theta_0(0) = A_0 + D_0$ from this follows to

$$\varepsilon = \frac{N + 8f_1(N)}{9 + N + 8f_1(N)} \tag{B28}$$

with

$$f_1(N) = \frac{1 + x - x^3 - x^4}{1 + x^4} \qquad (x = e^{-N/3}) \tag{B29}$$

In the form $1/\Theta(N)$, one finally obtains

$$\boxed{\frac{1}{\Theta} = N + \frac{9N}{N + 8f_1(N)}} \tag{B30}$$

EFFICIENCY AND MEAN TEMPERATURE DIFFERENCE OF HEAT EXCHANGERS WITH ONE SHELL PASS AND EVEN NUMBERS OF TUBE PASSES

In 1965, Allan D. Kraus and Donald Q. Kern [K4] derived the following expression for the efficiency of heat exchangers with one shell pass (laterally mixed) and $n = 2m$ tube passes ($m = 1, 2, 3, \ldots$):

$$\epsilon = \frac{2}{1 + R + \dfrac{2}{n}\sqrt{1+(nR/2)^2}\,\coth\dfrac{\mathrm{NTU}\sqrt{1+(nR/2)^2}}{n} + \dfrac{2}{n}f(z)} \qquad (C1)$$

$$f(z) = \frac{jz^j + (j-2)z^{j-1} + (j-4)z^{j-2} + \ldots - (j-4)z^2 - (j-2)z - j}{1 + z + z^2 + z^3 + \ldots + z^{j-1} + z^j}$$

$$(C2)$$

$$z = \exp(2\mathrm{NTU}/n) \qquad j = (n-2)/2 \; (= m-1)$$

With $Y = \mathrm{NTU}$ and $R = X/Y$ (Y is the tubeside NTU), the generalization of the equations for $2m = 2$ and 4 passes from Table 2.3 ($2m, m, m$) is:

$$\frac{1}{\Theta} = \varphi(Z_m) + \varphi(Y) - \varphi(Y/m) + \frac{X + (Y/m) - Z_m}{2} \tag{C3}$$

$$Z_m = \sqrt{X^2 + (Y/m)^2} \qquad \epsilon = Y\Theta \qquad \varphi(x) = x/(1 - e^{-x}) \tag{C4}$$

In the following, it will be shown that the efficiencies calculated from eqs. (C1, C2) [K4] and from eqs. (C3, C4) are, in fact, the same. To do so, we first rewrite eq. (C1), with $(1/\Theta) = Y/\epsilon$:

$$(1/\Theta)_{KK} = \frac{X + Y}{2} + \frac{Z_m}{2} \coth \frac{Z_m}{2} + \frac{Y}{2m} f(z) \qquad z = \exp(Y/m)$$

and with $x \coth x = \varphi(2x) - x$ into

$$(1/\Theta)_{KK} = \frac{X + Y}{2} + \varphi(Z_m) - \frac{Z_m}{2} + \frac{Y}{2m} f(z) \tag{C5}$$

Comparing eqs. (C5) with (C3) yields and expression for the equivalent of the function $f(z)$ in our shorter notation:

$$f_{\text{Ma}}(z) = m \left(\frac{2\varphi(Y)}{Y} - 1 \right) - \left(\frac{2\varphi(Y/m)}{Y/m} - 1 \right) \tag{C6}$$

or rewritten with the original meaning of $\varphi(x)$ and using z gives:

$$f_{\text{Ma}}(z) = m \frac{z^m + 1}{z^m - 1} - \frac{z + 1}{z - 1} \tag{C7}$$

$$f_{KK}(z) = \frac{\sum\limits_{k=0}^{j} (j - 2k)z^{j-k}}{\sum\limits_{k=0}^{j} z^k} \tag{C8}$$

Eq. (C8) is $f(z)$ from eq. (C2). With the sum of the finite geometric series

$$\sum_{k=0}^{j} z^k = \sum_{k=0}^{j} z^{j-k} = \frac{z^{j+1} - 1}{z - 1} = \frac{z^m - 1}{z - 1} = S_m \tag{C9}$$

the functions $f(z) \cdot S_m = F(z)$ from eqs. (C7) and (C8) become

$$F_{\text{Ma}}(z) = m \left(S_m + \frac{2}{z - 1} \right) - S_m \left(1 + \frac{2}{z - 1} \right) \tag{C10}$$

$$F_{KK}(z) = \sum_{k=0}^{j} (j - 2k)z^{j-k} \qquad \text{(C11)}$$

Eq. (C10) may be rearranged into

$$F_{Ma}(z) = (m - 1)S_m + \frac{2}{z - 1}(m - S_m) \qquad \text{(C12)}$$

Dividing the polynomial S_m by $(z - 1)$ results in

$$\frac{S_m}{z - 1} = \frac{m}{z - 1} + \sum_{k=0}^{j} k \cdot z^{j-k} \qquad \text{(C13)}$$

which, when introduced into eq. (C12), with $m - 1 = j$, leads to

$$F_{Ma}(z) = \sum_{k=0}^{j} k \cdot z^{j-k} - 2\sum_{k=0}^{j} k \cdot z^{j-k} = \sum_{k=0}^{j} (j - 2k)z^{j-k} \qquad \text{(C14)}$$

So $F_{Ma}(z)$ is identical to $F_{KK}(z)$. From this, it is clear now that the simple formulas (C3) and (C4) give, in fact, the same result as obtained from the relationship (C1) and (C2) as derived by Kraus and Kern.

SYMBOLS

A	m^2	transfer surface area
a	1	function of N_2/n (defined in eq. [2.189])
a_v	m^{-1}	volume specific surface area
B	m	width, breadth
B	1	function of N_1 and N_2/n (defined in eq. [2.191])
B	1	relative wall resistance (defined in eq. [3.93])
$\dot{B}$	kg/(mh)	falling film mass velocity (chapter 3, section 5)
b	1	$= 1 - a$ (defined in eq. [2.190], see also eq. [2.45])
b	m	gap width
C	1	**capacity flow rate ratio** $= \boxed{(\dot{M}c_p)_1/(\dot{M}c_p)_2}$ (with negative sign for counterflow)
c_D	1	drag coefficient
c_p, c_v	J/(kg K)	mass specific heat capacity (at constant pressure, and constant volume, respectively)
D	m	diameter
d	m	diameter
d_h	m	hydraulic diameter (see eq. [1.117])
F	1	correction factor for ΔT_{LM}, $\boxed{F = \Delta T_M/\Delta T_{LM},}$ **LMTD correction factor**
F, f	1	function symbols
Gz	1	Graetz number $= \rho c_p w d^2/(\lambda L) = RePr\, d/L$
g	1	function of N_2/n (defined in eq. [2.36])
H	m	height
H	J	enthalpy
$\dot{H}$	W	enthalpy rate
h	J/kg	mass specific enthalpy

J	1	number of cells of a cascade
i, j, k	1	number of a cell, counting index
K	W/K	$= kA$ (in chapter 1, section 1 only)
K	1	diameter ratio ($= d_i/d_0$) in annuli
k	W/(m^2 K)	**overall heat transfer coefficient**
L	m	length
Le	1	Lewis number $= \lambda/(\rho c_p \delta_{12})$
l	m	length
$l*$	m	characteristic length (defined in eq. [3.91] chapter 3, section 6)
M	kg	mass
$\dot{M}$	kg/s	mass flow rate
$\dot{m}$	kg/(m^{2}s)	mass flux
m	1	exponent, counting index
$m_{1,2}$	1	roots of a quadratic equation (in chapter 1, section 3)
N	1	**number of transfer units** $= \boxed{kA/(\dot{M}c_p),}$ **NTU**
Nu	1	Nusselt number $= \alpha d/\lambda$
n	1	number of tubes, rows, channels, turns, etc.
n_v	m^{-3}	volume specific number
Pe	1	Péclet number $= \rho c_p wd/\lambda = Re\ Pr$
Pr	1	Prandtl number $= \eta c_p/\lambda$
p	Pa	pressure
Q	J	heat
$\dot{Q}$	W	heat rate, heat duty
$\dot{q}$	W/m^2	heat flux
R	1	**capacity flow rate ratio** $= \boxed{(\dot{M}c_p)_1/(\dot{M}c_p)_2 = |C|}$
Re	1	Reynolds number $= \rho wd/\eta$
r_m	1	roots of characteristic equation, eigenvalue
r	m	radius coordinate
S	m^2	cross sectional area, flow cross section
s	m	tube pitch, wall thickness
T	K, °C	temperature
t	s	time
U	J	internal energy
u	m/s	flow velocity
V	m^3	volume
v	1	recirculation rate (see eq. [3.76] chapter 3, section 5)
W	J	work
$\dot{W}$	W	power (mechanic, electric, etc.)
w	m/s	velocity
X, Y, Z	1	nondimensional length coordinates ($= N_X, N_Y$)
x, y, z	m	length coordinates
Z	1	dimensionless mean overall heat transfer resistance
z	1	dimensionless local overall heat transfer resistance (see eq. [3.92], [3.97] in chapter 3, section 6)

GREEK SYMBOLS

α	W/(m^2 K)	**heat transfer coefficient**
β	m/s	mass transfer coefficient
β_c, β_d	1	velocity ratios (in eq. [2.222])
δ	1	relative wall thickness (in chapter 1, section 3)
δ	m	fin thickness (in chapter 3, section 7)
δ_{12}	m^2/s	diffusion coefficient
Δ	1	difference
E	1	unit matrix
ϵ	1	**normalized change in temperature (= efficiency)**
ζ	1	dimensionless length coordinate
η	Pa s	viscosity
η	1	fin efficiency
Θ	1	**normalized mean temperature difference**
Θ, ϑ	1	normalized temperatures
ϑ	K	temperature difference (in chapter 3, section 7)
$\varkappa$	1	$= (kA)_{LL}/(kA)$ in chapter 1, section 1, $= (kA)_i/(kA)_0$ in chapter 1, section 3
λ	W/(K m)	**thermal conductivity**
μ	1	function (defined in eq. [1.106])
ν	m^2/s	kinematic viscosity $= \eta/\rho$
ξ	1	friction factor
ξ	1	dimensionless length coordinate (in chapter 3, section 6 $= x/l^*$)
Π, Σ		product, sum
ρ	kg/m^3	density
σ	1	dimensionless film thickness (defined in eq. [3.96])
τ	1	dimensionless time
Φ, φ	1	function symbols
φ	$\circ$	angle
ψ	1	void fraction
ω	1	nondimensional stirrer power (defined in eq. [1.20])
ω	1	$= (1 + R^2)^{1/2}$ in eq. (2.88)

SUBSCRIPTS

0	at $z = 0$
1	stream 1, apparatus 1
2	stream 2, apparatus 2
∞	at steady state, far away from the surface
A	air
a	ambient, annulus
B	film mass velocity

b	boiling
bed	packed bed
c	collector, condensation, counter current
d	distributor
el	electric
F	feed
f	film
g	gaseous
I	initial
i	inner
i, j, k	number index
J	jacket
L	lid, loss
LM	logarithmic mean
l	liquid, at the position $z = l$
loc	local
M, m	mean
m, max	maximum
min	minimum
o	at the surface, outer
opt	optimum
P	pump, particle (in chapter 3, section 4)
p	parallel flow
R	recirculation (in chapter 3, section 5)
req	required
S	shell-side, stirrer, storage mass
s	solid
T	tube-side
V	vapor, vaporization
W	water, wall
X	X-stream (see Fig. 2.54)
Y	Y-stream (see Fig. 2.54)

SUPERSCRIPTS

.	(dot)	flow rate
′	(prime)	at the inlet (in chapter 1, section 3: first derivative)
″	(double prime)	outlet (in chapter 1, section 3: second derivative)
i		intermediate value
~	(tilde)	molar quantity
–	(bar)	integral average
(0)		estimated value, zeroth approximation

REFERENCES

[A1]   Abramowitz, M., and Irene A. Stegun, eds. *Handbook of Mathematical Functions*. New York: Dover Publications, 1965

[A2]   Afgan, N. H., and E. U. Schlünder. *Heat Exchangers, Design and Theory Sourcebook*. New York: McGraw-Hill, 1974.

[B1]   Bassiouny, M. K. *Experimentelle und theoretische Untersuchungen über Mengenstromverteilung, Druckverlust und Wärmeübergang in Plattenwärmeaustauschern*. Fortschr.-Ber.VDI Reihe 6 Nr.181. Düsseldorf: VDI-Verlag, 1985.

[B2]   Bassiouny, M. K., and H. Martin. "Flow Distribution and Pressure Drop in Plate Heat Exchangers." *Chem. Engng. Sci.* 39 (1984):693–704.

[B3]   Bassiouny, M. K., and H. Martin. "Temperature Distribution in a Four Channel Plate Heat Exchanger." *Heat Transfer Engng.* 6 (1985):58–72.

[B3a]   Bes, Th. "Method of Thermal Calculation for Rating Countercurrent and Cocurrent Spiral Heat Exchangers." *Wärme-u. Stoffübertr.* 22 (1988).

[B3b]   Bes, Th., and W. Roetzel. *Thermal Theory of Spiral Heat Exchangers*. (Manuscript soon to be published).

[B4]   Bier, W., and K. Schubert. *Herstellung von Mikrostrukturen mit großem Aspektverhältnis durch Präzisionszerspanung mit Formdiamanten*. KfK-Bericht Nr. 4363, Nuclear Research Center Karlsruhe (Feb. 1988).

[B5]   Bird, R. B., W. E. Stewart, and E. N. Lightfoot. *Transport Phenomena*. New York: J. Wiley & Sons, Inc., 1960.

[C1]   Chowdhury, K., H. Linkmeyer, M. K. Bassiouny, and H. Martin. "Analytical Studies on the Temperature Distribution in Spiral Plate Heat Exchangers." *Chem. Eng. Process.* 19 (1985):183–190.

[C2]   Chowdhury, K., M. K. Bassiouny, and H. Martin. "Druckverlust und Wärmeübergang bei Spiralwärmeübertragern." Wissenschaftliche Abschlußberichte 19. Internationales Seminar für Forschung und Lehre in Chemieingenieurwesen, Technischer und Physikalischer Chemie. Universität Karlsruhe, 1984.

[C3]   Cieslinski, P. J., and T. Bes. *Analytical Heat Transfer Studies in a Spiral Plate Heat Exchanger*. XVI Congr. of Refrigeration, Institut International du Froid (I.I.F.), Paris (1983):67–72.

[C4]   Cross, W. T., and C. Ramshaw. "Process Intensification: Laminar Flow Heat Transfer." *Chem. Eng. Res. Des.* 64 (1986):293–391.

[D1] Domingos, J. D. "Analysis of Complex Assemblies of Heat Exchangers." *Int. J. Heat Mass Transfer* 12 (1969):537–548.

[D2] Dunn, P., and D. A. Reay. *Heat pipes*, 3d ed. New York: Pergamon Press, 1982.

[F1] Fischer, F. K. "Mean Temperature Difference Correction in Multipass Exchangers." *Ind. Eng. Chem.* 30 (1938):377–383.

[G1] Gaddis, E. S., and E. U. Schlünder. "Temperaturverlauf und übertragbare Wärmemenge in Röhrenkesselapparaten mit Umlenkblechen." *Verfahrenstechnik* 9 (1975):617–621.

[G2] Gaddis, E. S. "Über die Berechnung des Austauscherwirkungsgrades und der mittleren Temperaturdifferenz in mehrgängigen Rohrbündelwärmeaustauschern mit Umlenkblechen." *Verfahrenstechnik* 12 (1978):145–149.

[G3] Gaddis, E. S., and E. U. Schlünder. "Temperature Distribution and Heat Exchange in Multipass Shell-and-Tube Exchangers with Baffles." *Heat Transfer Eng.* 1 (1979):43–52.

[G4] Gaddis, E. S., and A. Vogelpohl. "Über den Wirkungsgrad und die mittlere Temperaturdifferenz von Rohrbündelwärmeübertragern mit Umlenksegmenten und zwei rohrseitigen Gängen." *Chem. Eng. Process* 18 (1984):269–273.

[G5] Gardner, K. A. "Mean Temperature Difference in Unbalanced-Pass Exchangers." *Ind. Eng. Chem.* 33 (1941):1215–1223.

[G6] Gardner, K. A. "Mean Temperature Difference in Multipass Exchangers." *Ind. Eng. Chem.* 33 (1941):1495–1500.

[G7] Gardner, K. A., and J. Taborek. "Mean Temperature Difference: A Reappraisal." *AIChE J.* 23 (1977):777–786.

[G8] Gnielinski, V. "Berechnung des Wärme- und Stoffaustauschs in durchströmten ruhenden Schüttungen." *Verfahrenstechnik* 16 (1982):36–39.

[G9] Gregorig, R. *Wärmeaustausch und Wärmeaustauscher.* Frankfurt/Main: Sauerländer, 1973.

[H1] Hausen, H. "Erzeugung sehr tiefer Temperaturen." Volume 8 of *Handbuch der Kältetechnik* (R. Plank, ed.) Berlin: Springer Verlag, 1957.

[H2] Hausen, H. *Wärmeübertragung im Gegenstrom, Gleichstrom und Kreuzstrom.* Berlin: Springer, 1957.

[H3] *Heat Exchanger Design Handbook* (HEDH) (E. U. Schlünder, ed.) Washington, D.C.: Hemisphere Publ. Co., 1983.

[K1] Kays, W. M., and A. L. London. *Compact Heat Exchangers.* 3d ed. New York: McGraw-Hill, 1984.

[K2] Kern, D. Q., and A. D. Kraus. *Extended Surface Heat Transfer* New York: McGraw-Hill, 1972.

[K3] Klapper, K., G. Linde, and L. Müller. "SO_2 und NO_x- Ein Gesamtverfahren zur Rauchgasreinigung." *Chemie-Anlagen + Verfahren* May (1985).

[K4] Kraus, A. D., and D. Q. Kern. "The Effectiveness of Heat Exchangers with One Shell Pass and Even Numbers of Tube Passses." *ASME* paper 65-HT-18 (1965).

[M1] Marktübersicht "Wärmeaustauscher" in *Chemie-Ing.-Techn.* 56, no. 6 (1984):A282–A295; and 60, no. 3 (1988):A177–A189.

[M2] Martin, H. "Low Peclet Number Particle-to-Fluid Heat and Mass Transfer in Packed Beds." *Chem. Engng. Sci.* 33 (1978):913–919.

[M3] Martin, H. "Dimensionierung von Wärmeaustauschern mit senkrecht stehenden, innen mit kondensierendem Dampf beheizten Rohren." *Verfahrenstechnik* 16 (1982):757–761.

[M4] Mollekopf, N., and D. U. Ringer. "Multistream Heat Exchangers—Types, Capabilities and Limits." 18. Int. Centre Heat Mass Transfer Symp. Heat a. Mass Transfer in Cryoengineering a. Refrigeration, Dubrovnik, Yugoslavia, Sept. 1986.

[N1] Nusselt, W. "Die Oberflächenkondensation des Wasserdampfes." *Z. VDI* 60 (1916):541–546; 569–575.

[N2] Nusselt, W. "Eine neue Formel für den Wärmedurchgang im Kreuzstrom." *Techn. Mechanik Thermodyn.* 1 (1930): 417–422.

[P1] Paikert, P. "Entwicklungstendenzen auf dem Gebiet der Wärmeaustauscher." *Chemie-Ing.-Techn. 60*, no. 3 (1988):A166–A175.

[P2] Paikert, P. Personal communication.

[P3]   Pignotti, A. "Matrix Formalism for Complex Heat Exchangers." *J. Heat Transfer* 106 (1984):352–360.

[R1]   Reid, R. C., J. M. Prausnitz, and T. K. Sherwood. *The Properties of Gases and Liquids*. 3d ed. New York: McGraw-Hill, 1977.

[R2]   Roetzel, W., and F. J. L. Nicole. "Mean Temperature Difference for Heat Exchanger Design—A General Approximate Explicit Equation." *J. Heat Transfer* 97 (1975):5–8.

[R3]   Roetzel, W. "Thermische Berechnung von dreigängigen Rohrbündelwärmeübertragern mit zwei Gegenstromdurchgängen gleicher Größe." *Wärme-u. Stoffübertragung* 22 (1988):3–11.

[R4]   Roetzel, W. "Thermische Berechnung von Wärmeübertragersystemen mit umlaufendem Wärmeträger." *Brennstoff, Wärme, Kraft* 42 (1990).

[S1]   Schedwill, H. *Thermische Auslegung von Kreuzstromwärmeaustauschern*. Fortschr.-Ber. VDI-Z. Reihe 6 Nr. 19. Düsseldorf: VDI-Verlag, 1968.

[S2]   Schlünder, E. U. *Einführung in die Stoffübertragung*. Stuttgart: Thieme-Verlag, 1984.

[S3]   Schlünder, E. U., and F. Thurner: *Destillation, Absorption, Extraktion*. Stuttgart: Thieme-Verlag, 1986.

[S4]   Schlünder, E. U. "On the Mechanism of Mass Transfer in Heterogeneous Systems—In Particular in Fixed Beds, Fluidized Beds and on Bubble Trays." *Chem. Engng. Sci.* 32 (1977):845–851.

[S5]   Schmidt, E. *Einführung in die Technische Thermodynamik*. Berlin: Springer, 1963.

[S6]   Schmidt, Th.E. "Der Wärmeübergang an Rippenrohre und die Berechnung von Rohrbündel-Wärmeaustauschern." *Kältetechnik* 15 (1963):98–102; 370–378.

[S7]   Scholz, W. H. "Gewickelte Rohrwärmeaustauscher." *Linde-Berichte aus Technik und Wissenschaft* 33 (1973):34–39.

[S8]   Shah, R. K., and A. L. London. *Laminar Flow Forced Convection in Ducts*. New York: Academic Press, 1978.

[S9]   Shah, R. K., and S. G. Kandlikar. "The Influence of the Number of Thermal Plates on Plate Heat Exchanger Performance." Proceedings of the Eighth National Heat and Mass Transfer Conference, Andhra University, Andhra, India, Dec. 1985.

[U1]   Underwood, A. J. V. *J. Int. Petroleum Tech.* 20 (1934):145.

[V1]   *VDI-Wärmeatlas*. 6. Auflage Düsseldorf: VDI-Verlag, 1991.

[Z1]   Zaleski, T. "A General Mathematical Model of Parallel-Flow, Multichannel Heat Exchangers and Analysis of its Properties." *Chem. Eng. Sci.* 39 (1984):1251–1260.